Advances in Industrial Control

Advances in Industrial Control is a series of monographs and contributed titles focusing on the applications of advanced and novel control methods within applied settings. This series has worldwide distribution to engineers, researchers and libraries.

The series promotes the exchange of information between academia and industry, to which end the books all demonstrate some theoretical aspect of an advanced or new control method and show how it can be applied either in a pilot plant or in some real industrial situation. The books are distinguished by the combination of the type of theory used and the type of application exemplified. Note that "industrial" here has a very broad interpretation; it applies not merely to the processes employed in industrial plants but to systems such as avionics and automotive brakes and drivetrain. This series complements the theoretical and more mathematical approach of Communications and Control Engineering.

Indexed by SCOPUS and Engineering Index.

Proposals for this series, composed of a proposal form (please ask the in-house editor below), a draft Contents, at least two sample chapters and an author cv (with a synopsis of the whole project, if possible) can be submitted to either of the:

Series Editor

Professor **Michael J. Grimble**:
Department of Electronic and Electrical Engineering, Royal College Building, 204 George Street, Glasgow G1 1XW, United Kingdom;
e-mail: m.j.grimble@strath.ac.uk

or the

In-house Editor

Mr. **Oliver Jackson**:
Springer London, 4 Crinan Street, London, N1 9XW, United Kingdom;
e-mail: oliver.jackson@springer.com

Proposals are peer-reviewed.

Publishing Ethics

Researchers should conduct their research from research proposal to publication in line with best practices and codes of conduct of relevant professional bodies and/or national and international regulatory bodies. For more details on individual ethics matters please see: https://www.springer.com/gp/authors-editors/journal-author/journal-author-helpdesk/publishing-ethics/14214

Giulio Ferro · Riccardo Minciardi · Luca Parodi · Michela Robba

Optimization of Electric-Vehicle Charging

Scheduling and Planning Problems

Springer

Giulio Ferro
DIBRIS
University of Genoa
Genoa, Italy

Riccardo Minciardi
DIBRIS
University of Genoa
Genoa, Italy

Luca Parodi
DIBRIS
University of Genoa
Genoa, Italy

Michela Robba
DIBRIS
University of Genoa
Genoa, Italy

ISSN 1430-9491 ISSN 2193-1577 (electronic)
Advances in Industrial Control
ISBN 978-3-031-61916-8 ISBN 978-3-031-61917-5 (eBook)
https://doi.org/10.1007/978-3-031-61917-5

Mathematics Subject Classification: 93-XX, 90-XX

This Springer imprint is published by the registered company Springer Nature Switzerland AG
The registered company address is: Gewerbestrasse 11, 6330 Cham, Switzerland

To the memory of my beloved parents, Maria and Mario.

Riccardo Minciardi

To my beautiful little girl Alice.

Michela Robba

To Lillia and Ermanno, for their support all over these years.

Giulio Ferro

To my family. You are my inspiration to accomplish something meaningful every day.

Luca Parodi

A special thanks to our mentor Riccardo for his precious advice and help in our research career.

Series Editor's Foreword

Control engineering is viewed rather differently by researchers and those who must implement and maintain control systems. Researchers develop general algorithms with a strong underlying mathematical basis, whilst engineers have more immediate concerns over the limits of equipment, quality of control, safety and security, and plant downtime. The series *Advances in Industrial Control* attempts to bridge this divide and hopes to encourage the adoption of advanced control techniques when they are likely to be beneficial.

The rapid development of new control theory and technology has an impact on all areas of engineering. This monograph series focuses on applications, which are important since it is the challenges from industry that often stimulate the development of new control paradigms. A focus on applications is desirable if the different aspects of the "control design" problem are to be explored with the same dedication that "control synthesis" problems have received. The series provides an opportunity for researchers to present new work on industrial control and applications. It raises awareness of the substantial benefits that advanced control can provide whilst tempering its enthusiasm with a discussion of the challenges that can arise.

This monograph is concerned with the optimization of electric vehicle charging stations. The authors have produced this text at a very opportune time given the enormous interest in electric vehicles (EVs). They have an extensive record of publications in renewables, electric vehicles, and power systems. It is difficult to overstate the importance of optimized charging stations in the total electric vehicle network.

Chapter 1 considers practical aspects of the EV problem covering the technology involved. It also discusses the integration of EVs and smart grids in the transport system. Chapter 2 covers the modelling and simulation of electric vehicles for optimal planning and management of the system. The chapter begins with the modelling of batteries, and the need for battery management systems. Chapter 3 considers optimal charging in smart grids and explains the need for good forecasting of load demands given the renewable power available. The case study provided involves a mix of traditional and renewable power sources and four vehicles representing the EV demand.

Chapter 4 looks at a discrete-event approach for aperiodic scheduling in single-socket charging stations. The operational management of a microgrid is involved and an optimization problem is established. The battery state-of-charge is important in charging and in the energy management of electric vehicles whilst in use. An optimization problem is proposed with a solution methodology. A case study illustrates the ideas for different scenarios. Chapter 5 considers the optimal scheduling problem of charging stations with multiple sockets.

Chapter 6 considers the optimal scheduling of electric buses in a depot. This is one of the main applications for EVs. The optimization problem is defined and again a case study is included. Since such vehicles follow a given route, there are good opportunities for optimization that may not apply in, for example, private cars. Chapter 7 turns to the formalization of the joint traffic and energy demand assignment conditions in the presence of electric vehicles. Traffic network modelling is included.

Chapter 8, on planning for EVs, considers joint traffic assignment and the sizing of charging stations. The optimal sizing and siting of charging stations are important, and optimization is an obvious tool to use. Chapter 9 involves planning for the optimal placement of charging stations in an electrical distribution network. The case study involves a mixed-integer nonlinear optimization problem that has well-established software available.

The authors have valuable experience in the total systems aspects of the EV charging problem, and this text mainly addresses the system-level aspects of the optimization problem rather than the lower level feedback control problems. The monograph should spur interest in this exciting EVs research area, which is of considerable current interest.

Glasgow, UK
March 2024

Michael J. Grimble

Preface

The progressive shift from traditional vehicles to electric vehicles (EVs) is considered one of the key measures to achieve the objective of a significant reduction in the emission of pollutants, especially in urban areas. EVs will be widely used in a not-so-futuristic vision, and new technologies will be present for charging stations, batteries, and vehicles. Moreover, EVs will act as distributed energy resources since they can provide regulation services and power supply, thus enabling the vehicle-to-grid (V2G) operation or the vehicle-to-building (V2B) operation. In this way, they would also absorb excess electricity production from renewable sources and return it to the grid in periods of peak loads. The number of EVs and charging stations (CSs) is increased in the last years, but, unfortunately, wide usage of EVs may cause technical problems to the electrical grid (i.e., instability due to intermittent distributed loads), inefficiencies in the charging process (i.e., lower power capacity and longer recharging times), long queues, and bad use of CSs. Moreover, it is necessary to integrate the transportation and the electrical networks to plan the CS installation over the territory, the schedule of vehicles, and the optimal use of CSs.

In this book, attention is focused on applying control and optimization methods and approaches to energy systems in which EVs are present, with specific reference to planning and scheduling decision problems. In particular, this book aims to provide models and methods for the optimal management of EVs through an interdisciplinary approach that brings together knowledge from sectors of transportation, manufacturing, and smart grids.

The book is intended both for researchers and practitioners in the field and for graduated and Ph.D. students.

The optimal management of EVs and charging stations (CSs) is a decision problem across different disciplines: transportation and logistics, scheduling, and smart grids. In fact, in smart grids, energy production and storage systems are usually scheduled by an energy management system (EMS) to minimize costs, power losses, and CO_2 emissions while satisfying energy demands. When CSs are connected to a smart grid, EVs served by CSs represent an additional load to the power system to be satisfied and an additional storage system in the case vehicle-to-grid technology is enabled. However, the load generated by EVs is deferrable. It can be thought of as a process in which there are machines (CSs) that serve customers/products (EVs) based on release time, due date, deadline, and energy request, as it happens in manufacturing systems.

For this reason, in this book, firstly, attention is focused on applying the same approaches usually adopted for scheduling in manufacturing systems to the case of EVs. Usually, the scheduling of EVs in a smart grid is found through the formalization of a discrete-time optimization problem in which fossil fuel production plants, storage systems, and renewables are considered to satisfy the electrical load of the grid. The discrete-time formalization can use forecasting for renewables and loads without data elaboration. On the other side, a huge number of decision variables are present, making the optimization problem hard to solve through commercial optimization tools. For this reason, this book explains an alternative method for the optimal schedule of EVs: discrete-event formalization.

However, when dealing with EVs, it is necessary to consider other issues that influence the scheduling, i.e., the transportation demand assessment, the optimal location of charging stations, and the assignment of users to CSs. For this reason, the second part of the book is devoted to planning problems and charging stations' placement. It will be seen that it is necessary to consider the transportation network and the traffic assignment of both electric and conventional vehicles in order to assess energy demand and assign traffic. Finally, it will describe how territorial characteristics and the electrical grid may influence the decision for CSs' location and sizing.

The book is organized as follows. Chapter 1 introduces smart grids and the characteristics of EVs and CSs. The role of modeling, simulation, and optimization is then highlighted in Chap. 2, in which state of the art and current challenges will be described.

From Chaps. 3–6 the book is focused on the optimal scheduling of electric vehicles with different possible options in the problem formalization: discrete-time and discrete-event optimization, and aperiodic and periodic decision models. The different approaches are described in detail to give the basis for the development of models for the scheduling of EVs. In particular, Chaps. 3 and 4 focus attention on the scheduling of EVs through a discrete-time and a discrete-event formalization, respectively, in the case of an aperiodic framework. Chapters 5 and 6 are dedicated to discrete-event formalizations for aperiodic and periodic frameworks, respectively. Examples and applications to real case studies are reported too.

The remaining chapters are dedicated to planning problems. In particular, Chap. 7 is focused on user equilibrium models for energy demand assessment and traffic assignment. Chapter 8 is dedicated to a bi-level approach that is characterized by stochastic user equilibrium conditions at the lower level. At the higher level, the sizing and location of CSs are formalized. Finally, in Chap. 9, the exact placement on a distribution grid of CSs is discussed, which has, as input, the results of the analysis reported in Chaps. 7 and 8, and provides the optimal results for location and sizing of CSs, taking into account the electrical constraints of a smart grid and its EMS.

Conclusions and future developments are drawn in Chap. 10. An Appendix is introduced to remind concepts of optimization and modeling.

It is due to point out that the work in this book has been expanded by and shares some parts with the thesis of Luca Parodi.[1] In particular, the first two chapters share some of the main discussions about the main characteristics of EVs and the related equipment, as well as the main advancements in the state of the art. Chapters 3 and 4 reproduce the same approach adopted in the fourth and fifth chapters of the thesis and Chap. 5 proposes an extended version of Sect. 5.5 of the thesis.

Genoa, Italy

Giulio Ferro
Riccardo Minciardi
Luca Parodi
Michela Robba

Acknowledgments The authors thank the AMT (Azienda Mobilità e Trasporti) personnel for the sharing of data and information about the EBs' fleets. The work has been supported by the EU Inc. BIPE Programme of the University of Genoa.

[1] © L. Parodi. Reproduced from L. Parodi "Optimization of Electric-Vehicle Charging: Scheduling and Planning Problems", Ph.D. Thesis, 2023. https://hdl.handle.net/11567/1118383.

Contents

Abbreviations

AC	Alternating Current
BMS	Battery Management System
CCS	Combined Charging System
CHP	Combined Heat and Power
CSs	Charging Station
DBs	Diesel Buses
DC	Direct Current
DR	Demand Response
EBs	Electric Buses
EMS	Energy Management System
EOL	End of Life
EVCS	Electric Vehicle Charging Station
EVs	Electric Vehicles
EVSE	Electric Vehicle Supply Equipment
KERS	Kinetic Energy Recovery System
LCA	Life Cycle Assessment
Li-Ion	Lithium-Ion
MILP	Mixed-Integer Linear Program
MINLP	Mixed-Integer Nonlinear Problem
PV	Photovoltaic
SOC	State-of-Charge
SOH	State-of-Health
SUE	Stochastic User Equilibrium
TTW	Tank-to-Wheel
UE	User Equilibrium
V2G	Vehicle-to-Grid
WT	Wind Turbine
WTT	Well-to-Tank

Chapter 1
Introduction

Sustainability is one of the most widespread words in recent years. It can be considered in different frameworks as the economic, the social, or the environmental one. Deepening the latter, it is necessary to focus on pollution. Among the numerous sources of pollution, transportation represents a considerable contribution. This means that through an improvement in this field, it is possible to affect global pollution positively. One of the most promising technologies is electric vehicles (EVs) for private and public (usually electric buses, EBs) transportation which is the actual focus of this book. In this first chapter, an introduction to the various aspects relevant to EVs is presented.

1.1 Electric Vehicles: Technical Characteristics

Electric traction vehicles are moved only by an electric motor, without the presence of an electric or mechanical power generator powered by an engine of another type. The general scheme of this type of vehicle (Fig. 1.1) can be represented by one or more electric motors and an energy storage system.

Besides, EVs have the advantage of not emitting harmful exhaust gases on site. However, to assess greenhouse gas emissions, it is necessary to consider the entire life cycle of the vehicle and all its phases, starting with the production and transport of electricity. Therefore, the emission analysis integrates two phases [1, 2]:

- the assessment of the well-to-tank (WTT) provides the emissions during the extraction and distribution of fuel and the production and distribution of energy;
- the tank-to-wheel (TTW) evaluation, on the other hand, measures the emissions during the use of the vehicle and, therefore, those measured at the tailpipe.

Of course, indirect emissions (WTT) depend on the primary source, the production method, and the distribution route. Furthermore, they are closely linked to the energy

G. Ferro et al., *Optimization of Electric-Vehicle Charging*, Advances in Industrial Control, https://doi.org/10.1007/978-3-031-61917-5_1

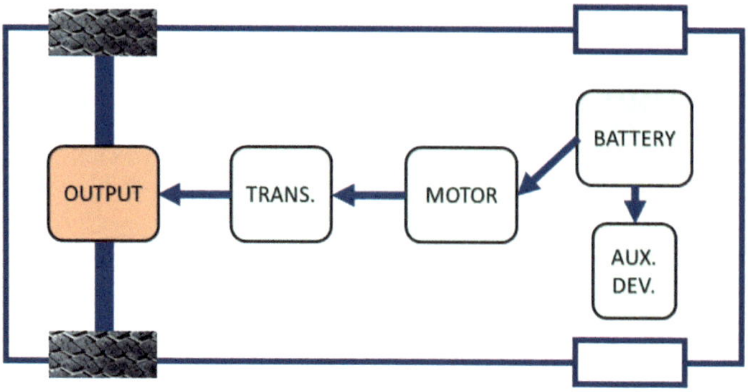

Fig. 1.1 Configuration of an EV

mix at the local level. In particular, if the energy mix is mainly based on producing energy from renewable sources (RES), this type of emission is shallow.

The assessment of direct emissions (TTW) is complex, and their extent depends on the type of vehicle (fuel used, abatement devices, etc.) and the context (driving conditions, traffic situation, use of auxiliaries, load profile, etc.). If we consider a "pure electric" vehicle, this type of emission is zero.

It can be concluded that "pure electric" vehicles can be considered the most effective solution for reducing greenhouse gas emissions in the transportation sector if they are powered by renewable sources.

The authors in [3] compare conventional diesel buses (DBs) and EBs regarding polluting emissions and consumption of oil and fossil fuels. The fossil fuels considered in this analysis are coal, oil, and natural gas. They show how most of an EB's energy demand and emissions are passed upstream in the production phase. In all the speed and traffic conditions taken into consideration, purely electric buses are advantageous from the point of view of both oil and fossil fuel consumption. Considering that the cost of primary energy sources is constantly growing, reducing such consumption would lead to significant savings in terms of operating costs.

Furthermore, they highlight how oil is reduced more than fossil fuels because there is still a significant consumption of coal for the production of upstream energy. Regarding carbon dioxide emissions, although EBs can be considered zero emissions in the TTW phase, thermoelectric power plants produce a substantial amount of CO_2. Therefore, carbon dioxide emissions are reduced by 19–24% in EBs compared to DBs. This highlights the need to increase the contribution of renewable sources to electricity production to make the penetration of EBs advantageous.

Despite the advantages from an environmental point of view, EBs, and in general all EVs, present a series of criticalities mainly linked to their propulsion system. The latter is characterized by an electric motor to which the necessary power is provided by batteries that must be recharged in specific stations. The system, therefore, requires an infrastructure that allows the charging process of the batteries which must be appropriately sized to guarantee the vehicle the necessary autonomy which, in any case, is reduced compared to that of a conventional vehicle.

The whole commercial, marketing, and economic history of the EVs' production have been associated with the electric battery autonomy problems in terms of feasible distance and hour duration. They are used to store electrical energy that the electric motor uses to power the vehicle. Unfortunately, current battery systems tend to be heavy and costly, so the priority is to improve battery systems to make electrification as accessible as possible.

Moreover, the battery represents the most expensive component of an EV, since its cost is from 25 to 50% of the total, depending of which technology is used [4, 5].

1.2 Batteries

Battery improvements are moving fast in order to make EVs competitive with respect to conventional vehicles: for instance, production costs of the Li-ion batteries steeply decreased in the last decades reaching a minimum price of about 150$/kWh in 2022, and it is expected to decrease in the next years [5].

EVs can be equipped with various battery-manufacturing technologies. As referenced in [6], Table 1.1 details the technologies currently favored by the manufacturing industry.

In the current framework, Li-ion batteries emerge as the predominant technology in EVs [7], owing to their high energy density. This characteristic enables the creation of battery variants with reduced weight and dimensions, all at competitive prices.

Lithium-ion batteries also have some negative aspects relevant to their functioning at extremely low or high temperatures or in case of a too low voltage or an overload; in this case, it is possible to see a performance decrease or even malfunctioning. Therefore, nowadays, the aim is to develop a new battery system capable of operating even in particularly unfavorable situations and equipping lithium-ion batteries with a management system to be controlled and managed effectively.

Table 1.1 Main EV battery types: advantages and disadvantages

Technology	Advantages	Disadvantages
Lead–acid	Cheap manufacturing	Low energy density; presence of pollutants
Nickel–cadmium	High lifespan	Presence of pollutants
Nickel–metal hydride	Lack of memory effect	Presence of pollutants
Lithium-ion	High energy density	High costs; limited life cycle
Lithium-ion–polymer	High energy density	High costs; functional instability in case of overload or extreme discharge
Sodium–nickel-chloride	High energy density	High-temperature operations

For safety and reliability to be guaranteed, this system has to impose that the battery operates within an operating range constrained by temperature and voltage limits.

In the realm of EVs, batteries typically exhibit relatively low capacity and voltage. These batteries comprise individual cells integrated into modules, with an EV's battery pack housing one or more modules as per its needs. Consequently, the battery pack comprises numerous individual cells. To manage this array of cells effectively, an EV necessitates a battery management system (BMS) encompassing sensors, actuators, and controllers.

The main tasks of the BMS in vehicles are to (a) protect cells and battery packs from damage; (b) keep voltage and temperature in specific operation ranges, ensuring safety and extending battery life; (c) make batteries operate such as they meet the vehicle's requirements.

Moreover, the battery is not symmetrical between the charge and discharge phases: this can be derived from the fact that the equivalent internal resistance has two different values depending on the current verse due to the different chemical reactions involved in the two different phases. Therefore, a battery discharged in a certain amount of time will require more time to be recharged. In addition, the current that the battery can sustain is low: combined with the low maximum cell voltage, the incoming power will also assume low values. As a result, it will take quite a long time (hours) to recharge the battery fully. This is inevitably a disadvantage compared to the traditional vehicle, both petrol and diesel, which instead has relatively low "recharge" times (to fill up the car, it takes one to two minutes at most).

It is essential to consider that the capacity of a lithium-ion battery does not have a constant value but is reduced due to aging processes. The "end of life" (EOL) of a battery is defined as an 80% remaining capacity (or as a doubled internal resistance, depending on what occurs before).

The condition of the EOL is essential for sizing the battery capacity: there is, in fact, the necessity to consider a 20% bound as a reserve. The battery is used under strict constraints to limit and decelerate aging, and the "state-of-charge" (SOC) must fall within a predefined range. The lower threshold also referred to as the "safety limit" should never be exceeded to ensure battery reliability and performance stability. As soon as the SOC level drops below this threshold, the EV is recharged so that the battery is protected from excessive exhausting and complete exhaustion does not stop the vehicle.

On the other hand, the upper limit is set by the maximum level of charge status compatible with full charging power. If this threshold is exceeded, the charging power is reduced. The fast charging causes a further reduction in usable capacity due to voltage limits. At high values of the charging state, the current with which the vehicle is recharged must be reduced not to exceed the battery's upper voltage limit. This effect increases with the battery's aging due to the increase in internal resistance. Reducing the current increases the charging time, going against the concept of "fast charging". The upper region of the SOC cannot, therefore, be loaded too quickly.

Even with all these measures, the degradation of the battery must be taken into account with an initial oversizing of the battery capacity itself.

The ideal battery can be charged to a 100% SOC with maximum current charge, and its capacity remains constant.

In any case, the sizing of the battery is the result of a compromise between the required range and the weight and cost of the battery pack and, therefore, the vehicle.

Another very useful characteristic that must be taken into account is the life cycle assessment (LCA). This parameter takes into account the relationship between input components and output factors and can determine the resulting environmental impact.

The main takeaway from the LCA of batteries is that, for each EV, the ecological impact and longevity of the vehicle rely on every stage of the production, use, and EOL of the vehicle.

An integral part of a comprehensive LCA is the recycling process.

Materials like cobalt, nickel, and manganese are forwarded for refining [8, 9], while lithium and rare earth elements end up in slag. Of course, this final point is a matter of current discussions and could represent a key improvement in the sustainability of battery adoption.

Despite the advantages from an environmental point of view, all EVs present a series of criticalities. Among them, a critical point is the one related to the charging process which usually has to be carried out in specific infrastructures.

1.3 Charging Stations: Technical Characteristics

Before presenting the technical characteristics of the charging station, a brief introduction on the main differences between the charging processes.

In general, there are four possible "speeds", namely

- slow charging;
- fast charging;
- rapid charging;
- ultra-rapid charging.

Slow charging rates range from 2.3 to 3 kW, obtaining a full charge in around 10–14 h. Typically, this charging is considered for overnight charging.

The fast-charging range is from 7 to 22 kW and can be typically found in public areas such as parking lots. It is suitable for users who have few hours to charge the vehicle.

Rapid charging refers to those chargers that allow (AC or DC) charges at 43 and 50 kW.

Ultra-rapid chargers can deliver 100, 150, or 350 kW. These are only DC chargers.

The well-known company Tesla uses 150 ultra-rapid chargers, which have the infrastructure for charging EVs. This chapter does not focus on this particular case, but we will describe the common characteristics identified by international standards.

Describing the technical characteristics of the charging facilities, it is necessary to cite the IEC 62196 [10]. It is an international standard by the International Electrotechnical Commission (IEC) related to electrical connectors for EVs.

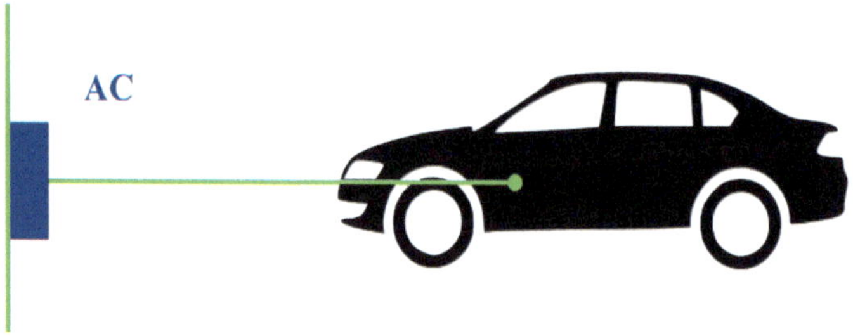

Fig. 1.2 Connection in “Mode 1” configuration

In particular, it is based on the IEC 61851 [11], another international standard in which general characteristics relevant to the chargers are established. IEC 62196 comprises information relevant to the different couplers distinguished by the considered current flow (AC or DC). According to the standard, each of the four connector types must include control signaling, allowing the control of local charging. Moreover, it is specified all connectors can be converted by means of specific adapters, although the charging mode is necessarily affected.

1.4 Charging Modes

The standard applies to EV charging appliances characterized by an operating voltage lower than 690 V (1550 V) in AC (DC). In particular, this standard’s reference is represented by the charging modes defined in the IEC 61851, summarized in the next subsections.

1.4.1 Mode 1

Mode 1 (Fig. 1.2) is characterized by a direct and passive AC connection of the EV, a limit of 16 A. The connection does not include extra control pins. Connection to ground and ground fault protection must be provided to the EV by the EV supply equipment (EVSE). Since not all household installations have the necessary grounding, Mode 1 is illegal in certain nations.

1.4.2 Mode 2

Mode 2 (Fig. 1.3) connects the EV directly and semi-actively to the AC mains with a maximum current of 32 A. In particular, it is important to specify that this connection

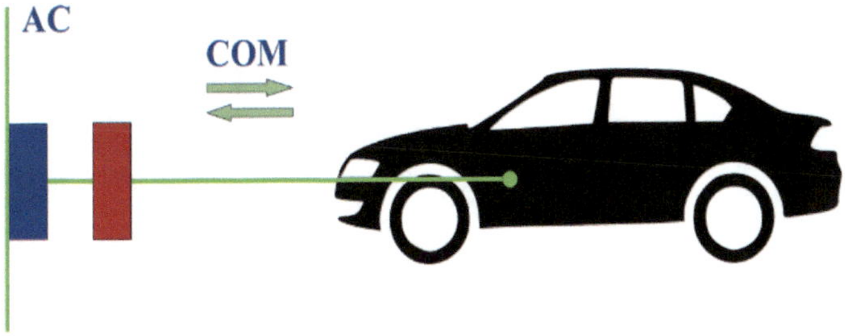

Fig. 1.3 Connection in "Mode 2" configuration

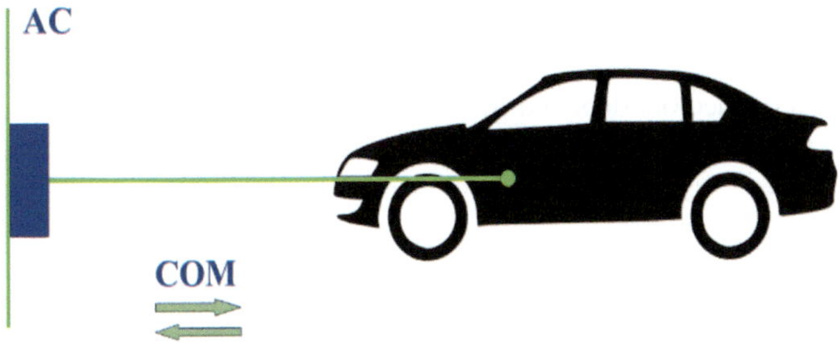

Fig. 1.4 Connection in "Mode 3" configuration

is direct and passive while the connection between the EVSE and the EV is active, with the addition of the control pilot to the passive components.

1.4.3 Mode 3

Mode 3 (Fig. 1.4) is characterized by an active connection between the EV and the EVSE. It may include a compulsorily captive cable with extra conductors. To enable the active connection of the charging supply, proper communication over the control pilot is necessary.

1.4.4 Mode 4

As seen for Mode 3, also Mode 4 (Fig. 1.5) comprehends an active connection. It is the only connection mode providing DC current (600 V). The maximum current that can be provided is 400 A and both grounding and control pilot must be included in the charging supply. The EVSE directly rectifies AC mains power into DC charging power.

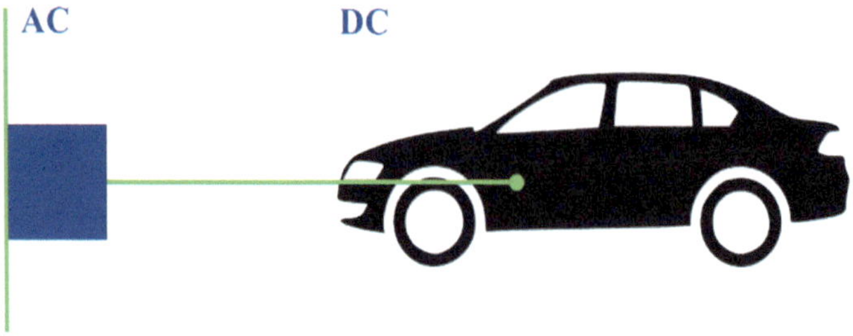

Fig. 1.5 Connection in "Mode 4" configuration

1.5 Connectors

The standard connectors depicted in IEC 62196 are specifically designed for EVs. The list of plug types includes four different connectors characterized by different specifics.

1.5.1 Type 1

Type 1 (Fig. 1.6) connector comes from the North American standard SAE J1772/2009. The standard provides the fundamental physical, electrical, communication, and performance specifications for the conductive charge system and coupler of an electric vehicle.

Type 1 is characterized by five pins, from the greater to the smaller: AC (2 pins), ground (1 pin), proximity detection, and control pin.

Fig. 1.6 Type 1 (SAE J1772) connector

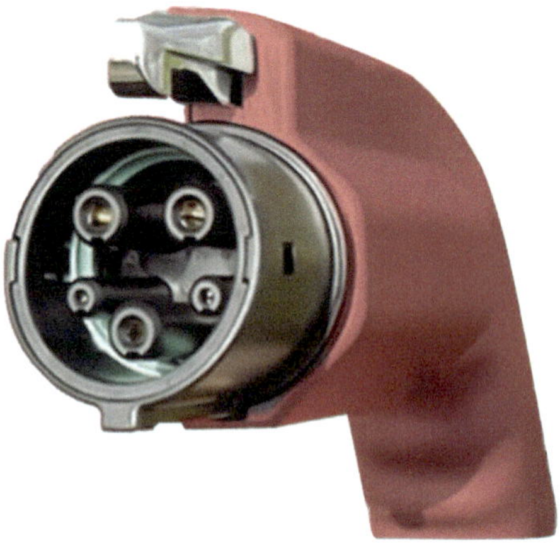

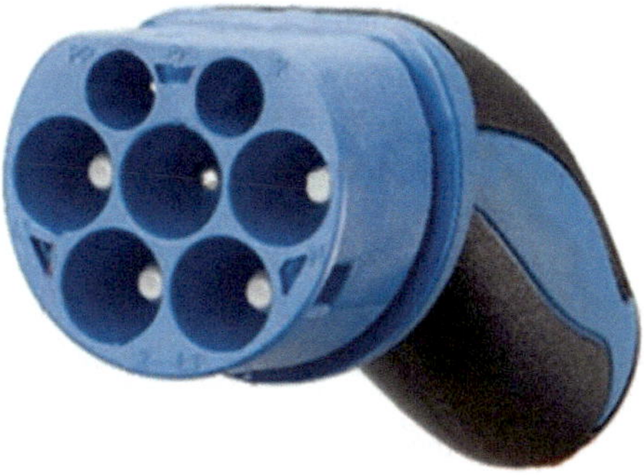

Fig. 1.7 Type 2 (Mennekes) connector

1.5.2 *Type 2*

The second type of connector (Fig. 1.7), named "Type 2", but commonly referred to as Mennekes, is the most spread in Europe since it has been selected as the official charging plug within the European Union. The power the connector can provide is between 3 and 120 kW. Both alternating current (AC) and direct current (DC) are admitted for this connector.

1.5.3 *Type 3*

The EV Plug Alliance connector (Fig. 1.8) is defined by IEC 62196 as "Type 3". It is derived from a previous model by the Italian company Scame and can provide a maximum of 32 A with a 3-phase system.

Fig. 1.8 Type 3 connector

1.5.4 Type 4

Type 4, connector is reported in Fig. 1.9. This connector is mainly known by its trade name (CHAdeMO) and represents a quick charging method delivering up to 62.5 kW.

Moreover, it is necessary to introduce the combined charging system (CCS) which adds DC charging to the Type 1 and Type 2 connectors. These are commonly known as Combo 1 or Combo 2 connectors (Fig. 1.10). Generally, Combo 2 is the most used.

Note that there are other available connectors. An example is given by the American Tesla, which has developed its own connectors and its charging stations, but also provides an adapter to Type 2, CHAdeMO, and CCS Combo 2 connectors.

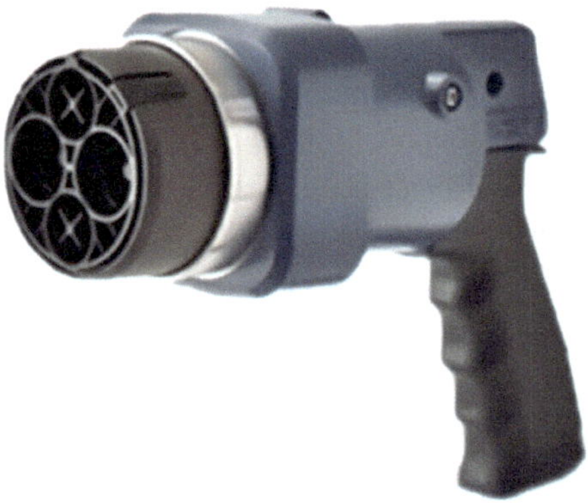

Fig. 1.9 Type 4 (CHAdeMO) connector

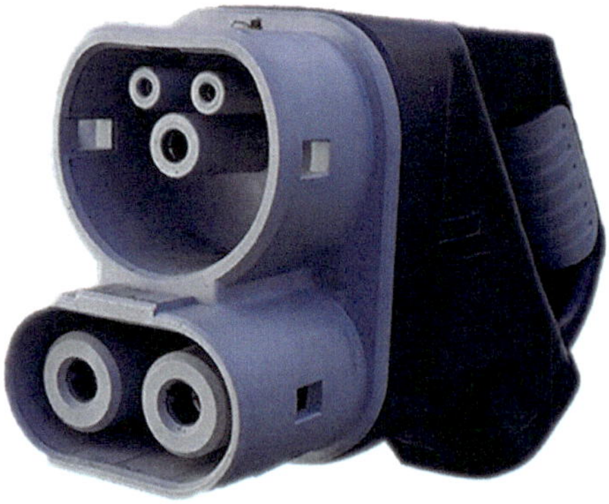

Fig. 1.10 CCS Combo 2 connector

1.6 Vehicle-to-Grid and Smart Charging

1.6.1 Vehicle-to-Grid (V2G)

This term describes a system in which EVs can communicate with the power grid to help balance it during the electric peak demands, working as temporary electrical energy storage [12].

Since about 95% of cars are parked [13], the batteries in EVs could be used to let electricity flow from their battery to the distribution network and be charged before a certain deadline. In particular, this would happen in response to peak load needs. The extra nighttime generation can then be absorbed, while these cars are recharged at less expensive rates during off-peak hours. Here, power is buffered by using the automobiles as a distributed battery storage system.

V2G vehicles, by utilizing "valley-filling" (charging during off-peak hours) and "peak-shaving" (sending power back to the grid during peak hours), offer a valuable resource for load balancing (refer to Fig. 1.11). They are utilized for regulation services to maintain voltage and frequency stability, while also providing spinning reserves to meet sudden demands for power.

Furthermore, V2G EVs have the potential to stabilize the intermittency of renewable power and incentivize the penetration of renewables in the electric market.

1.6.2 Smart Charging (SC)

The common definition of smart charging refers to a charging process of EVs that can be scheduled to follow a certain policy. Usually, these policies depend on both the grid and vehicle owner's needs. Participating in these programs is usually remunerated with some benefits which can be monetary and non-monetary.

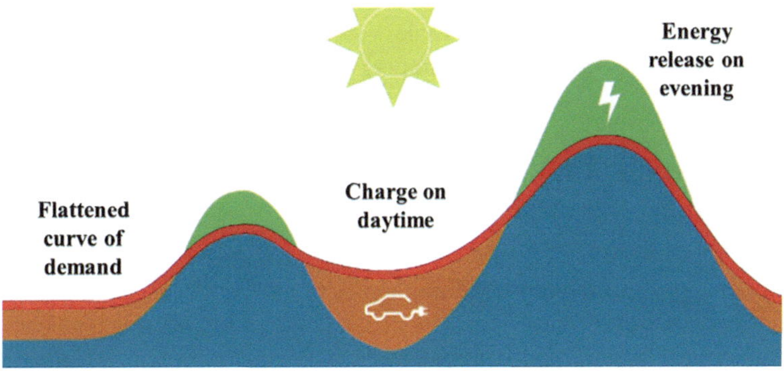

Fig. 1.11 Vehicle-to-grid mechanism

The fundamental issue to be considered is the effects on the power demand produced by the introduction of EVs. An uncontrolled charging process depending only on EV owners' habits may produce disruptive peaks in the power demand. A plausible case would see a large amount of EVs starting their charge during the evening when the consumption is already particularly high. This would obviously lead to a strong unbalance in the power grid.

Nevertheless, utilities can match supply and demand by properly managing EVs' charging. For example, EVs' charging would be a perfect fit for the integration of PV plants while charging in the central hours of the day when the production is high.

These SC strategies can be very effective in reducing emissions and lower electricity rates, and providing grid services.

A SC infrastructure must have the following fundamental features:

- the available capacity must be proportionally distributed (***Load Balancing***) over all the active charging stations while ensuring the optimal charging to all EVs at a specific location, within the limits of the charging stations' capacity;
- ***HUB/satellite connection***, which can collect in one time the data coming from multiple charging facilities;
- ***peak shaving***, which automatically reduces the consumption or pauses a charging session, until enough power becomes available;
- ***drivers' priority***: the EVs should be charged and be available whenever the driver needs them;
- ***time-of-use pricing consciousness***: prices must be defined by taking into account supply and demand profiles in order to avoid peaks.

Finally, some advantages of the SC application are

- due to abundant solar electricity, wholesale markets are experiencing a decline in mid-day power prices, thereby influencing retail prices for customers;
- smart chargers have the capability to adapt charging rates according to fluctuations in power supply, ensuring efficient energy consumption;
- peak mitigation would be observed in the electric grid, optimizing its functionality and stability;
- solar power systems selling wholesale would witness heightened revenues from increased daytime demand for their energy output.

1.7 Integration of EVs and Smart Grids/Transportation Systems

Smart grids are one of the most exciting topics of the last decade. They are usually medium and small-size power networks characterized by sensors that give information about the system and control power management better. Usually, a smart grid integrates different technologies such as renewables and traditional plants. EVs recently hold an essential role since one of their main characteristics is the large amount of power they involve. This means that they must be carefully considered when modeling such smart grids. The work in [14] provides an illustrative case, presenting a novel load management solution designed to coordinate the charging activities of numerous EVs within a smart grid framework. They consider cost minimization by incorporating time-varying market energy prices and preferred charging time zones for EV owners.

EVs can also be used for frequency regulation, as presented by [15]. The authors propose a real-time decision-making model to perform a dynamic control of the power exchange between a fleet of EVs and the power grid in order to provide a frequency regulation service.

The presence of EVs is also significant when considering demand response (DR) policies. DR stands for the variation of power demand as an ancillary service to the grid, a power decrease, and an increase if necessary. DR being a service to the grid, it is remunerated and could represent an economic resource [16]. An example of an application is given by [17]. The authors present an EMS that manages the EVs' charge–discharge plan and PV curtailment to reduce the operational cost while preserving the EVs' usage for driving. Another work focusing on demand response policies is [18], where a mathematical formulation includes innovative distribution companies' owners of renewable power plants and EVs' parking lots.

A pretty novel approach is presented in [19], where the authors present a model for managing the charging process of slow-charge EVs and fast-charge EVs while providing dynamic regulation of the grid. They consider that some slow-charge EVs can provide enough power to reduce the peak caused by the fast-charge EV.

The authors in [20] investigate the optimal energy management problem of a microgrid with EVs while minimizing costs. The EVs' owners are incentivized to participate in DR programs as a flexible load since it brings profit for EVs and microgrid owners. Even if the most significant number of studies focuses on discrete-time models, there are some applications of discrete-event models to determine the optimal scheduling of some EVs. The authors in [21] propose a discrete-event approach for an optimal scheduling problem. In [22] the authors present a model for a charging station with multiple sockets adopting a discrete-event formulation for the optimal charging schedule. In both studies, the results show that the discrete-event approach can reduce the number of variables with an admissible loss in the solution's optimality.

A relevant result is proposed by [23, 24]. The authors present a new approach concerning the coupling of the electric and the traffic networks to determine the optimal planning of some charging stations.

References

1. Woo JR, Choi H, Ahn J (2017) Well-to-wheel analysis of greenhouse gas emissions for electric vehicles based on electricity generation mix: a global perspective. Transp Res Part D: Transp Environ 51:340–350. https://doi.org/10.1016/j.trd.2017.01.005

2. Raykin L, Roorda MJ, MacLean HL (2012) Impacts of driving patterns on tank-to-wheel energy use of plug-in hybrid electric vehicles. Transp Res Part D: Transp Environ 17(3):243–250. https://doi.org/10.1016/j.trd.2011.12.002
3. Zhou B et al (2016) Real-world performance of battery electric buses and their life-cycle benefits with respect to energy consumption and carbon dioxide emissions. Energy 96(2016):603–613. https://doi.org/10.1016/j.energy.2015.12.041
4. Nykvist B, Nilsson M (2015) Rapidly falling costs of battery packs for electric vehicles. Nat Climate Change 5(4). https://doi.org/10.1038/nclimate2564
5. Chen Z, Guo N, Li X, Shen J, Xiao R, Li S (2017) Battery pack grouping and capacity improvement for electric vehicles based on a genetic algorithm. Energies 10(4). https://doi.org/10.3390/en10040439
6. Manzetti S, Mariasiu F (2015) Electric vehicle battery technologies: from present state to future systems. Renew Sustain Energy Rev 51. https://doi.org/10.1016/j.rser.2015.07.010
7. Iclodean C, Varga B, Burnete N, Cimerdean D, Jurchiş B (2017) Comparison of different battery types for electric vehicles. IOP Conf Ser: Mater Sci Eng. https://doi.org/10.1088/1757-899X/252/1/012058
8. Mansur MB, Guimarães AS, Petraniková M (2022) An overview on the recovery of cobalt from end-of-life lithium ion batteries. Miner Process Extr Metall Rev 43(4):489–509
9. Zhu L, Chen M (2020) Development of a two-stage pyrolysis process for the end-of-life nickel cobalt manganese lithium battery recycling from electric vehicles. Sustainability 12(21):Art. no. 21. https://doi.org/10.3390/su12219164
10. IEC (2011) IEC 62196-2 ed1.0: Plugs, socket-outlets, vehicle connectors and vehicle inlets—Conductive charging of electric vehicles. http://webstore.iec.ch/webstore/webstore.nsf/Artnum_PK/45683
11. IEC (2010) IEC 61851-1 ed2.0: Electric vehicle conductive charging system—Part 1: General requirements. http://webstore.iec.ch/webstore/webstore.nsf/Artnum_PK/44636
12. Kempton W, Tomić J (2005) Vehicle-to-grid power implementation: from stabilizing the grid to supporting large-scale renewable energy. J Power Sources 144(1):280–294. https://doi.org/10.1016/j.jpowsour.2004.12.022
13. Mwasilu F, Justo JJ, Kim EK, Do TD, Jung JW (2014) Electric vehicles and smart grid interaction: a review on vehicle to grid and renewable energy sources integration. Renew Sustain Energy Rev 34:501–516. https://doi.org/10.1016/j.rser.2014.03.031
14. Deilami S, Masoum AS, Moses PS, Masoum MAS (2011) Real-time coordination of plug-in electric vehicle charging in smart grids to minimize power losses and improve voltage profile. IEEE Trans Smart Grid 2(3). https://doi.org/10.1109/TSG.2011.2159816
15. Islam MM, Zhong X, Sun Z, Xiong H, Hu W (2019) Real-time frequency regulation using aggregated electric vehicles in smart grid. Comput Ind Eng 134:11–26. https://doi.org/10.1016/j.cie.2019.05.025
16. Shaukat N et al (2017) A survey on electric vehicle transportation within smart grid system. Renew Sustain Energy Rev 81:1329–1349. https://doi.org/10.1016/j.rser.2017.05.092
17. Kikusato H et al (2019) Electric vehicle charge-discharge management for utilization of photovoltaic by coordination between home and grid energy management systems. IEEE Trans Smart Grid 10(3):3186–3197. https://doi.org/10.1109/TSG.2018.2820026
18. Sadati SMB, Moshtagh J, Shafie-khah M, Catalão JPS (2018) Smart distribution system operational scheduling considering electric vehicle parking lot and demand response programs. Electr Power Syst Res 160:404–418. https://doi.org/10.1016/j.epsr.2018.02.019
19. Khalkhali H, Hosseinian SH (2020) Multi-class EV charging and performance-based regulation service in a residential smart parking lot. Sustain Energy Grids Netw 22:100354. https://doi.org/10.1016/j.segan.2020.100354
20. Aliasghari P, Mohammadi-Ivatloo B, Alipour M, Abapour M, Zare K (2018) Optimal scheduling of plug-in electric vehicles and renewable micro-grid in energy and reserve markets considering demand response program. J Clean Prod 186:293–303. https://doi.org/10.1016/j.jclepro.2018.03.058

21. Ferro G, Laureri F, Minciardi R, Robba M (2019) A predictive discrete event approach for the optimal charging of electric vehicles in microgrids. Control Eng Pract 86:11–23. https://doi.org/10.1016/j.conengprac.2019.02.004
22. Ferro G, Minciardi R, Parodi L, Robba M (2020) Discrete event optimization of a vehicle charging station with multiple sockets. Discrete Event Dyn Syst: Theory Appl. https://doi.org/10.1007/s10626-020-00330-0
23. Ferro G, Minciardi R, Robba M (2020) A user equilibrium model for electric vehicles: joint traffic and energy demand assignment. Energy 117299. https://doi.org/10.1016/j.energy.2020.117299
24. Ferro G, Minciardi R, Parodi L, Robba M (2021) Optimal planning of charging stations in coupled transportation and power networks based on user equilibrium conditions. IEEE Trans Autom Sci Eng 19(1):48–59

Chapter 2
Modeling and Simulation for the Optimal Planning and Management of Electric Vehicles: State of the Art and Challenges

2.1 Introduction

This chapter discusses the role of modeling and simulation for the optimal planning and management of electric vehicles (EVs). In particular, a crucial point for EVs is related to the modeling and simulation of batteries, which are very important to estimate the driving range of a vehicle but also for EVs' integration in smart grids, microgrids and buildings as storage systems (both for smart charging and vehicle-to-grid operations). For these reasons, in this chapter, attention is focused on batteries' models and management systems. Moreover, it is also discussed how these models are used in controllers for charging stations, and optimization models for smart grids and buildings. Indeed, when EVs are integrated into general power and energy systems, it is also necessary to model all the other elements that contribute to the overall power balance. These elements correspond to the electrical and thermal networks, the production plants, and components like transformers, inverters, etc., at different spatial and temporal scales. This chapter is a general introduction to the topics of modeling and simulation. Then, in the next chapters, attention will be focused on the mathematical formalizations necessary for the optimal planning, scheduling, and control. Specific attention will be paid to the different possible representations of battery models, on how to deal with nonlinear behavior and bidirectional flows, both in the case of discrete-time and discrete-event optimization problems.

G. Ferro et al., *Optimization of Electric-Vehicle Charging*, Advances in Industrial Control, https://doi.org/10.1007/978-3-031-61917-5_2

2.2 Storage Systems Modeling and Management

2.2.1 Modeling Batteries

Electrical storage systems play a crucial role in managing smart grids and EVs. As regards smart grids, they can store energy and use it in periods in which the power from the external grid has a high price or when demand response programs are needed. For EVs, when the vehicle-to-grid (V2G) mode is possible, the vehicle's battery works like a storage system in a smart grid, and thus it has cycles of charging and discharging. Moreover, in the case of EVs, it is also necessary to model the charge/discharge of the battery over a path that depends on many factors such as territorial characteristics, vehicle mass, velocity, and acceleration. Battery models aim to predict the operation of a battery based on discharge rate, charge rate, battery age, battery type, and temperature, and, often, when accurate models are considered, result in complex nonlinear functions [1]. There are many methods of modeling battery operation, and the major categories are mathematical models, electrochemical models, and electrical equivalent circuit networks [2–5].

Electrochemical models are based on physics and are represented in general by nonlinear equations, which may be simplified under many different approximations [2]. A popular model is the lumped parameter, which assumes a uniform spatial distribution of chemical products and is described by a small set of differential equations. Indeed, this model is considered too simplified for modern Li-ion cells because it doesn't well represent the complex electrochemical processes. Instead, the porous electrode theory is considered more reliable for Li-ion cells because it is possible to include mass transport and diffusion, side reactions, temperature, ion distribution, and aging. Finally, there are those models that use electrical components to model the battery behavior. These circuital models (of which the most common are the Thévenin models) can represent time-dependent effects and generally include a series of resistors and capacitors (see Fig. 2.1). During discharging, the chemical products near the cathode and anode are consumed, and voltage gradually decreases, while during charging products diffuse from the battery's body to the anode and cathode.

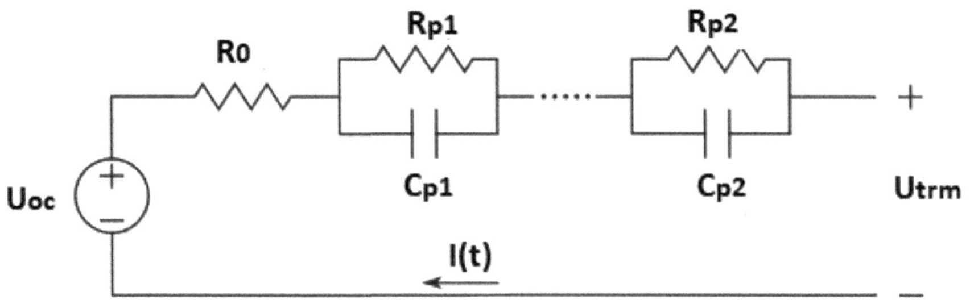

Fig. 2.1 Dynamic RC battery model

Figure 2.1 illustrates a circuit model in which Uoc represents the battery's open-circuit voltage, while R0 refers to the internal Ohmic resistance that is intrinsic to both the circuits and electrodes. Additionally, Rp1 and Cp1 explain the battery dynamics associated with reaction kinetics and surface effects on the electrodes caused by the formation of double layers. Conversely, Rp2 and Cp2 embody the slower dynamics representing diffusion processes within the electrolyte and active components.

Circuit-equivalent models can vary in complexity, depending on the number of elements included, such as resistors and capacitors. Moreover, these models can be refined by including various internal states of the battery and considering temperature-induced effects.

When dealing with EVs, battery modeling is crucial in determining charging times at stations, anticipating the range of EVs, and enhancing performance when merged with smart grids, microgrids, and charging facilities. It is worth mentioning that [3] provides a thorough study of battery modeling, particularly on EVs. Within this discourse, the primary categories include mathematical models, electrochemical models, and electrical equivalent circuit networks. The literature also presents hybrid model types, such as analytical-electrochemical and thermal models specific to batteries.

When considering battery modeling for EV applications, several critical aspects require consideration, including SOC estimation. A practical approach favors a rough estimate instead of an elaborate battery model due to computational efficiency. It is imperative to have swift models that can meet the real-time operational demands of the battery management system (BMS).

The model must also effectively handle the high discharge rates peculiar to EVs.

2.2.2 Battery Management Systems

A BMS is a tool for the monitoring and control of charging and discharging in a battery in order to optimally manage it, to prevent damages, and to guarantee a long life of the component [6]. In particular, it is important to reduce overcharging, to avoid the battery being empty, to analyze and store the SOC, to maintain low supply voltage, and to receive and export data from/to external platforms. Figure 2.2 reports the architecture of a classical BMS.

One of the most significant functions of the BMS is the estimation of SOC [3]. In this case, simple models generally guarantee fast operation and control of the battery. The techniques used are, for example, fuzzy-logic, least squares regression models, and Extended Kalman Filter. Another important parameter (more difficult to estimate and measure) is the state-of-health (SOH), i.e., the battery's gradual loss in terms of maximum capacity. In [7], the authors present an extensive review of battery modeling and state estimation approaches for advanced BMSs. The survey in [8] focuses on EVs' SOC estimation methods, noting that research in this field is

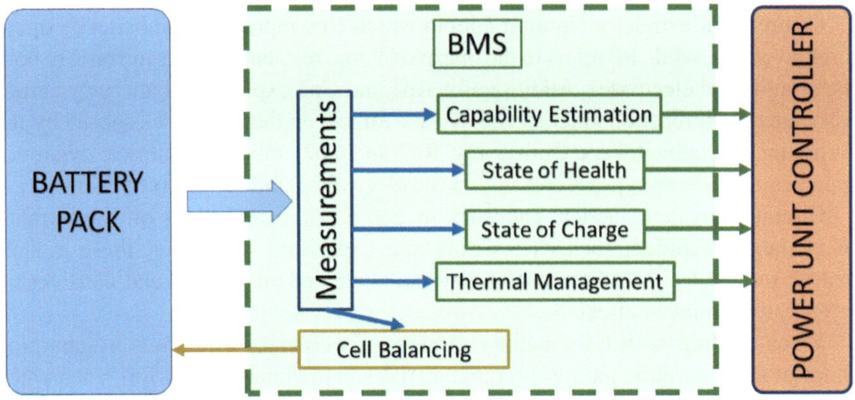

Fig. 2.2 BMS architecture

centered around control theory and intelligent algorithms. A review of the state-of-health (SOH) estimation is instead reported in [9], with specific reference to Li-ion batteries.

One of the most critical issues related to batteries, especially Li-ion batteries, is thermal management [10]. A BMS that ensures acceptable temperature levels of the battery pack has a significant effect on the system's performance, safety, and stability. Multi-physics-based thermal management systems for batteries are classified into three groups, namely air-cooled, liquid-cooled, and heat-pipe-cooled systems. An overview is presented in [11] for the specific topic of electric buses for public transport. BMSs are particularly important in this context because these buses have a shorter range than conventional internal combustion engines, require larger battery packs, and face practical obstacles such as long charging times and the high cost of large batteries.

In particular, the following issues are treated as a priority for the research in this field:

- development of technologies such as energy storage systems, powertrains, interleaving elements, and electric motors;
- development of tools and ICT-based automation systems for energy storage systems sizing, power/energy management, range remedy methods, charging/design/scheduling, etc.;
- modeling of charging demands and impact on power systems.

BMSs can also be seen as ICT platforms communicating with hardware in the field concerning a local component, which may include simulation and optimization models and can receive inputs from external platforms. The typical case is one of a storage system/vehicle connected to a microgrid or a building. In this case, the microgrid's management system must coordinate several components and provide reference working points to local controllers (i.e., BMS, building automation systems, and controllers for production plants). The BMS tries to follow such reference values but preserve at the same time the good health and operation of the component; this may result in a discrepancy between what the microgrid's controller asks and what the battery implements. For this reason, the modeling of storage systems is also very important for its inclusion in energy management systems for microgrids and building automation systems.

2.2.3 Modeling Energy Consumption over a Path

EVs' batteries have charging and discharging cycles during a trip. It is essential to model the state-of-charge over a path for several reasons: to quantify energy consumption to be used in optimal charging strategies, to develop optimization models for the optimal routing and charging of EVs, to design the size and the location of charging stations over a territory, and to control velocity and acceleration of an EV dynamically. The energy consumption over a path depends on the energy required to proceed on different slopes and the energy obtained by the kinetic energy recovery system (KERS) in the downward stroke and the braking phases. In the literature, there are many papers related to the motion of vehicles. In [12], a survey is presented that discusses existing mathematical models of EVs. The article presents both simple models and complex multi-body dynamic models, concentrating on their use in the design of controllers. In [13], a method is presented for optimizing the energy consumption of EVs through torque demand control. This approach includes a demand control method and a predictive controller model. It is important to note that there are some simplifications of vehicle dynamics for EVs' routing and charging, planning, and scheduling accepted in literature [14–16].

The force required for the propulsion is the summation of the acceleration force, the force necessary for the elevation of the bus F_{mg}, the force required to overcome the friction between the tires and the road F_{roll}, and the drag force F_{drag} (see Fig. 2.3) [14, 15]. The rolling resistance force F_{roll} is given by $F_{\mathrm{roll}} = C_r mg\cos(\theta)$, with θ the road gradient in radians or degrees, C_r the rolling resistance coefficient (function of tire material, structure, temperature and inflation pressure, road roughness, and material), m the vehicle mass, and g the gravity acceleration. The aerodynamic drag force F_{drag} is expressed as $F_{\mathrm{drag}} = \frac{\rho A C_d v^2}{2}$, with C_d being the aerodynamic drag coefficient, A the electric vehicle's frontal area, ρ the air density, and v the velocity. The grading resistance F_{mg} is given by $F_{\mathrm{mg}} = mg\sin(\theta)$.

The electric power consumption can be generally written as [16]

$$P_{\mathrm{el}} = \frac{v(t)}{\eta_{\mathrm{te}}\eta_e\eta_{\mathrm{in}}}\left(\psi m\frac{\mathrm{d}v(t)}{\mathrm{d}t} + C_r mg\cos(\theta) + mg\sin(\theta) + C_d\frac{\rho A v(t)^2}{2} + P_{\mathrm{HVAC}}\right) \tag{2.1}$$

where ψ is the mass factor for converting the rotational inertia of rotating components into translational mass, $v(t)$ the speed of the vehicle, η_e the electrical motor efficiency,

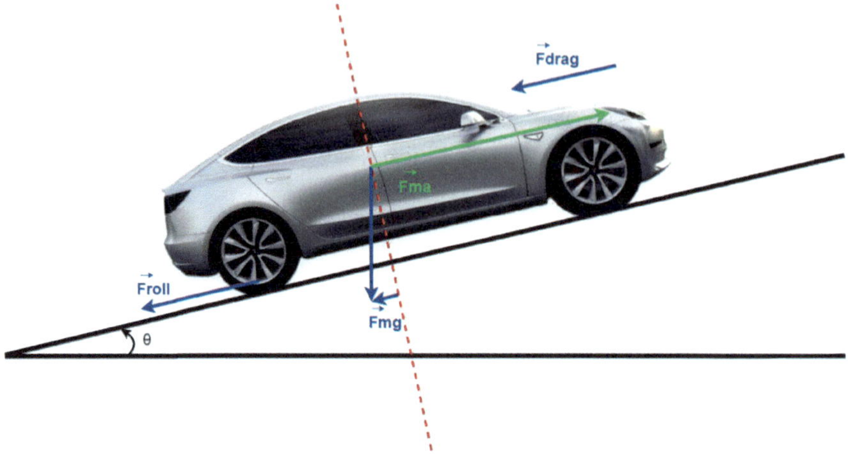

Fig. 2.3 The forces of a vehicle in motion

η_{in} the power converter efficiency, η_{te} the electric vehicle's transmission efficiency, and P_{HVAC} the internal power consumption of the vehicle (e.g., for air-conditioning).

The regenerative braking power P_{rb}, which can be partially (according to factor $K \in [0, 1]$) recovered and restored into the battery, can be modeled as

$$P_{\text{rb}} = Kv(t)\eta_{\text{te}}\eta_e\eta_{\text{in}}\left(\psi m_{\text{tot}}\frac{\mathrm{d}v(t)}{\mathrm{d}t} + C_r m_{\text{tot}} g \cos(\phi) + m_{\text{tot}} g \sin(\phi) + C_d \frac{\rho A v(t)^2}{2}\right). \tag{2.2}$$

It is important to note that there are other aspects that strongly influence the EVs' consumption. The first one is related to the number of start and stops over a path. Let $\boldsymbol{\Theta}$ be the number of start and stops over a path that can be estimated on the basis of traffic, length, and road characteristics. Figure 2.4 shows the start and stop for a path with $\boldsymbol{\Theta} = 3$ and constant velocity between start and stops. In this case, it is necessary to evaluate the additional contribution in consumption and recovery by the use of equations similar to (2.1) and (2.2) and multiplied by the number of start and stops $\boldsymbol{\Theta}$ over a path.

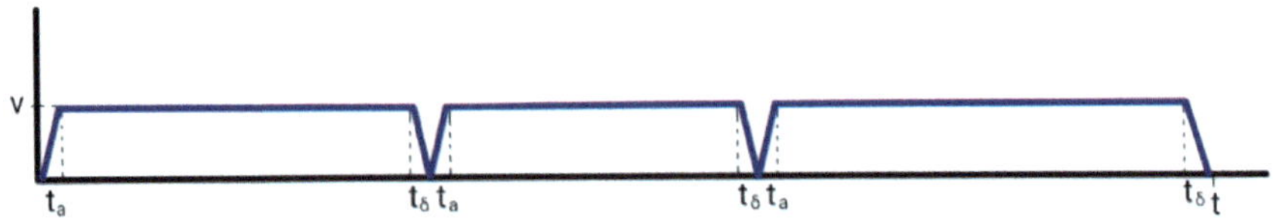

Fig. 2.4 Start and stops over a path with $\boldsymbol{\Theta} = 3$

2.3 Optimization Models for EVs' Planning and Scheduling in Smart Grids

Optimization models are widely used for the planning and management of EVs. In particular, the decision problems that should be faced are

- *Optimal schedule of the EVs' charging in buildings and microgrids.* In this case, we refer to decisions in the short term (day-ahead, intra-day). It is necessary to define how much power to provide, when, and at which vehicle, based on release time, due date, and deadline associated with each vehicle. Moreover, since the charging station is within an electrical network in which there are production plants, storage systems, and loads, it is necessary to jointly schedule the charging, the production and storage systems, and flexible loads.
- *Optimal sizing and siting of charging stations.* Decisions are taken in a long-term time horizon regarding installing plants and components. In particular, it is necessary to define the charging station, the size, the geographical location, and which bus of the electrical grid the charging station should be connected to.
- *Power management in charging stations.* This is the case in which we have multiple vehicles already connected to a socket in a charging park. It is necessary to set all vehicles as soon as possible but respect the constraint of the maximum power taken from the external grid. We refer here to short-term and real-time decisions.
- *Optimal routing and charging of electric vehicles.* This is the case in which EVs transport goods or persons but also have to recharge in their path (and decide in which charging station to recharge, how much, and which route to follow). We refer here to short-term decisions.

In this book, we refer to the first two classes of decision problems, which are detailed in the following.

From the existing literature, it is clear that the siting and sizing of charging stations (CSs) are crucial issues for the wide spreading of EVs. Several points should be taken into account:

- *Territorial constraints.* The characteristics of an area may influence the choice of the siting of CSs: (a) in some places, it is not possible to install CSs due to lack of space or limits given by regulation or risk assessment (natural areas, hydrogeological risk, etc.); (b) private citizens, commercial centers, public administration, and companies may have the willingness to install some CSs in priority areas.
- *Electrical grid management.* CSs are connected to the distribution grid, and EVs represent intermittent and significant loads. Moreover, when V2G capabilities are enabled, EVs can be treated as a particular type of storage system and may help manage the electrical grid. From a purely electrical grid point of view, EVs' charging and discharging are part of the overall power balance and optimal power flow. Typically, they are treated together with other loads and production systems.
- *Energy demand assessment.* The energy demand for charging is a crucial point that, for private transportation, depends on user choices, traffic models, and the characteristics of the transportation network. For public transportation (electrical

buses), energy demands for all possible paths can be estimated through a model of consumptions over the territory and the forecasting of the number of passengers and traffic modeling.

In the literature of planning problems, Pagany et al. 2018 [17], present a survey of spatial localization methodologies for the electric vehicle charging infrastructure to minimize costs while guaranteeing a high coverage of a charging station and trip length. User choices' modeling is present in various contributions in the literature. For example, in [18], a user equilibrium (UE) traffic assignment approach assesses traffic and energy demands in a transportation network. In contrast, in [19, 20], UE conditions are used as part of multilevel architectures in which they are considered as constraints of an optimization problem or part of an iterative procedure. Other articles extend these approaches to the case in which the electrical grid is modeled in detail [21–23] or stochastic equilibrium is considered [24].

Different from the sizing and siting of charging stations, which refer to long-term decision problems for installing technologies, in the short term (when all CSs are all defined over the territory), there is another significant decision problem: the optimal scheduling of electric vehicles' charging. This decision problem should be integrated with the optimal management of the system in which CSs are located (i.e., smart grids, microgrids, buildings), as represented in Fig. 2.5. EVs can fit this system since they represent not only a load that can be modulated but also a resource. In a not-so-futuristic vision, they can act as distributed energy resources since they can favor active short-response participation in storage resources by providing regulation services and power supply.

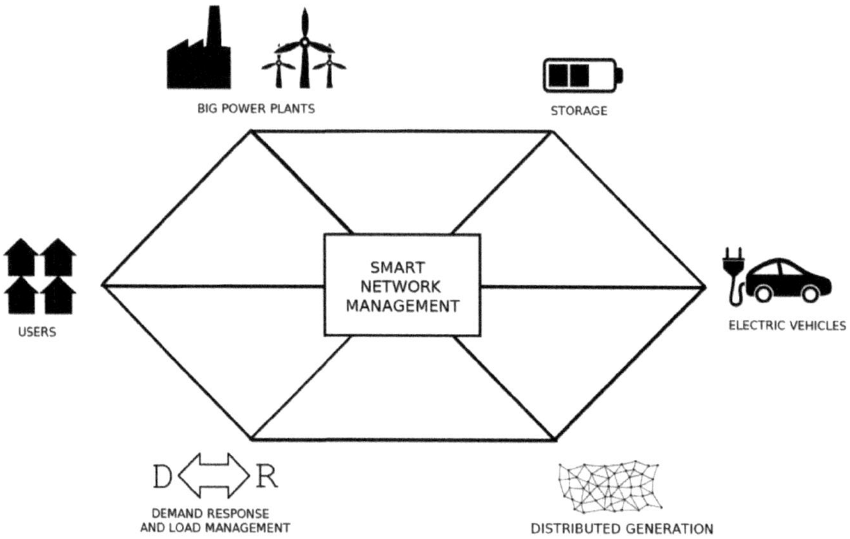

Fig. 2.5 A smart grid that includes EVs

Indeed, the advent of the smart grid era and the concomitant development of advanced communication and information infrastructures have highlighted the necessity for bidirectional communication, advanced metering infrastructure, and energy management systems (EMSs) for microgrids and districts. These systems are indispensable for the optimal scheduling of production systems, storage systems, and loads [25]. An energy management system (EMS) is a collection of information and communication technology (ICT) tools employed by operators of electric utility grids and microgrids to oversee, manage, and enhance the performance of power generation and distribution systems in their respective locales. Optimization models provide the "intelligence" for these EMSs, and a wide-ranging survey of microgrid control topics is presented in [26]. Various techniques are grouped into centralized, decentralized, distributed, and hierarchical frameworks. In [27], the paper reviews the management of loads in buildings under a demand response framework, with considerations for electric vehicles, renewable resources, storage systems, and automation for monitoring and control. In [28], the authors propose an optimal control problem for a storage system in a microgrid. In [29], implementing a multi-objective scenario-based day-ahead energy management system can lower operating costs and boost a microgrid's dependability by addressing electrical and thermal loads. In [30], the Savona Campus pilot site employs a comprehensive energy management system featuring diverse electrical models, electric vehicles, storage systems, combined heat and power (CHP) systems, and multiple objective functions. Wencong et al. [31] present a stochastic model for scheduling energy in a microgrid that accounts for intermittent energy sources like wind and solar, controllable loads such as EVs, and distributed generators such as micro gas turbines and diesel generators. They also consider the effect of distributed energy storage devices like battery banks. Meanwhile, You et al. [32] suggest a charging strategy for EV stations that operates within a dynamic electricity pricing environment. The scheduling problem is formulated as a constrained mixed-integer linear program (MILP) to capture the discrete nature of the battery states, specifically, charging, idle, and discharging. The authors in [33] present a model for the optimal scheduling of some EVs in a microgrid characterized by renewables, storage systems, and traditional plants. It is therefore clear that the main modeling needs concern:

- production systems (both renewables and fossil fuel plants);
- thermal and electrical storage systems;
- batteries of EVs in charging and discharging modes;
- charging stations;
- thermal and electrical distribution networks;
- performance indicators: costs, emissions, primary energy savings, etc.

As described in the following chapters, optimization problems can be found in the literature for discrete-time and discrete-event optimization problems.

In the former case, the required information and models are reported in Fig. 2.6. Firstly, it is necessary to obtain the following data required in each time interval of the optimization horizon:

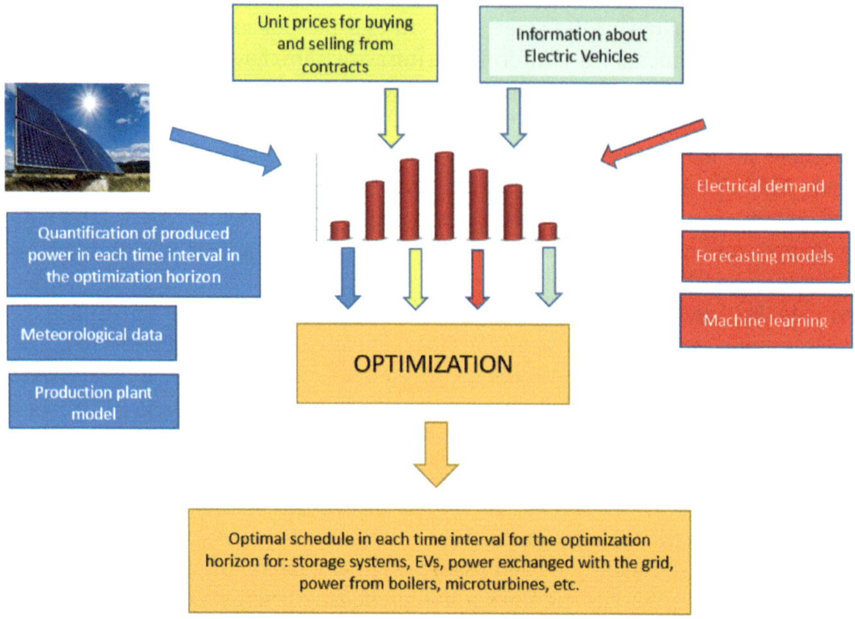

Fig. 2.6 Inputs and outputs for an optimization model with time discretization

- the prediction of power from renewable resources, which is calculated based on forecasted environmental parameters (temperature, solar radiation, wind velocity, etc.), and that can be derived by physically based models, data-driven models, and the model of the plants;
- energy contracts can be used to derive the prediction of unit costs and benefits;
- forecasts' overloads, which are generally derived by black-box models and the use of machine learning techniques;
- information about vehicles' arrival, energy request, and due date;
- initial state-of-charge of the batteries (measured in the field).

After all the inputs have been calculated, it is necessary to run the optimization model with the desired performance criteria (costs, emissions, primary energy savings, etc.) and the system model's constraints.

By running the optimization problem, one obtains the optimal schedule in each time interval for all production plants and components.

A similar approach can be done in the case of discrete-event optimization, in which time intervals are not present and there is a system change whenever an "event" occurs at a certain time. The main difference, in terms of information in input and output to/from the optimization problem, is that the EMS collects data in discrete time, but then it is required to fit them with an analytical function (as described in the following chapters). Moreover, all control and state variables are not discretized

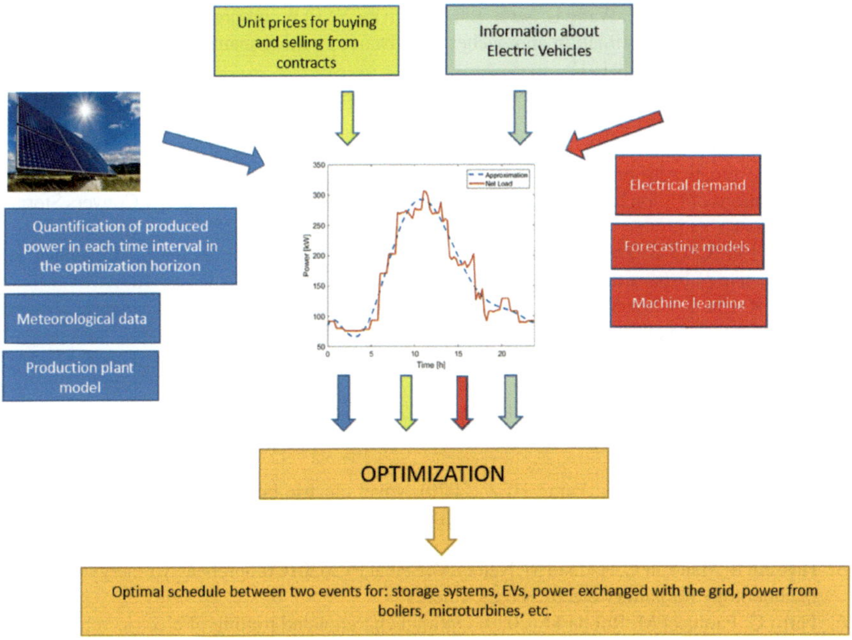

Fig. 2.7 Inputs and outputs for an optimization model in a discrete-event approach

in time. Figure 2.7 reports the inputs and outputs for an optimization problem in a discrete-event approach.

Some examples of a discrete-event approach are presented in [34, 35], where the authors use this approach in the optimal scheduling of some EVs' charging processes.

References

1. Adam SR, Reed GF, Karestes KRJ, Grainger BM, Smith ZT (2012) Survey of battery energy storage systems and modeling techniques. In: Proceedings of IEEE power and energy society general meeting
2. Aden S, Thanh Son D, John M (2014) A survey of mathematics-based equivalent-circuit and electrochemical battery models for hybrid and electric vehicles simulation. J Power Sources
3. Fotouhi, Auger D, Propp K, Longo S, Wild M (2016) A review on electric vehicle battery modelling: from lithium-ion toward lithium-sulphur. Renew Sustain Energy Rev 56:1008–1021
4. Jongerden MR, Haverkort BR (2008) Battery modeling. Centre for Telematics and Information Technology, University of Twente, TR-CTI
5. Shafiei, Momeni A, Williamson S (2011) Battery modeling approaches and management techniques for plug-in hybrid electric vehicles. In: Proceedings of the IEEE vehicle power and propulsion conference, Chicago, IL
6. Bergveld HJ, Kruijt WS, Notten PHL (2002) Battery management systems: design by modelling. Springer, p 192

7. Wang Y, Tian J, Sun Z, Wang L, Xu R, Li M, Chen Z (2020) A comprehensive review of battery modeling and state estimation approaches for advanced battery management systems. Renew Sustain Energy Rev 13:1–18
8. Zhang M, Fan X (2020) Review on the state of charge estimation methods for electric vehicle battery. World Electr Veh J 11(1):1–17
9. Kalita P, Garg A, Niu XD, Zhang XW, Peng X, Bhattacharjee D (2019) A review of state of health estimation of energy storage systems: challenges and possible solutions for futuristic applications of Li-ion battery packs in electric vehicles. J Electrochem Energy Convers Storage 16(4):1–12
10. Lin J, Liu X, Li S, Zhang C, Yang S (2021) A review on recent progress, challenges and perspective of battery thermal management system. Int J Heat Mass Transf 167:1–16
11. Deng R, Liu Y, Chen W, Liang H (2021) A survey on electric buses: energy storage, power management and charging scheduling. IEEE Trans Intell Transp Syst 22:9–22
12. Lekshmi S, Lal Priya PS (2019) Mathematical modeling of Electric vehicles—a survey. Control Eng Pract 92:1–17
13. He Z, Shi Q, Wei Y, Zheng J, Gao B, He L (2021) A Torque demand model predictive control approach for driving energy optimization of battery electric vehicle. IEEE Trans Veh Technol 70:3232–3242
14. El-Taweel NA, Zidan A, Farag HEZ (2020) Novel electric bus energy consumption model based on probabilistic synthetic speed profile integrated with HVAC. IEEE Trans Intell Transp Syst 24
15. Hussein B, Mansour C, Haddad M, Nemer M, Stabat P (2020) Comprehensive energy modelling methodology for battery electric buses. Energy
16. Ferro G, Paolucci M, Robba M (2020) Optimal charging and routing of electric vehicles with power constraints and time of use energy prices. IEEE Trans Veh Technol 69:14436–14447
17. Pagany R, Ramirez Camargo L, Dorner W (2018) A review of spatial localization methodologies for electrical vehicle charging infrastructure. Int J Sustain Transp 13:433–449
18. Ferro G, Minciardi R, Robba M (2020) A user equilibrium model for electric vehicles: joint traffic and energy demand assignment. Energy 198:1–10
19. Hosseini M, MirHassani SA (2015) Selecting optimal location for electric charging stations with queue. KSCE J Civ Eng 19:2271–2280
20. Xiang Y, Liu J, Li R, Li F, Gu C, Tang S (2016) Economic planning of electric vehicle charging stations considering traffic constraints and load profile templates. Appl Energy 178:647–659
21. Ferro G, Minciardi R, Parodi L, Robba M (2021) Optimal planning of charging stations in coupled transportation and power networks based on user equilibrium conditions. IEEE Trans Autom Sci Eng 1–12
22. Wang S, Dong ZY, Luo F, Meng K, Zhang Y (2018) Stochastic collaborative planning of electric vehicle charging stations and power distribution system. IEEE Trans Industr Inf 14(1):321–331
23. Wang, He F, Lin X, Shen Z, Li M (2019) Designing locations and capacities for charging stations to support intercity travel of electric vehicles: an expanded network approach. Transp Res Part C: Emerg Technol 102:210–232
24. Zhou Z, Sun H, Guo Q (2018) Stochastic user equilibrium in charging station selection based on discrete choice model. Presented at 2018 IEEE Power & Energy Society General Meeting (PESGM)
25. Zhou B, Li W, Chan KW, Cao Y, Kuang Y, Liu X, Wang X (2016) Smart home energy management systems: Concept, configurations, and scheduling strategies. Renew Sustain Energy Rev 61:30–40
26. Sen S, Kumar V (2018) Microgrid control: a comprehensive survey. Annu Rev Control 45:118–151
27. Badar, Anvari-Moghaddam A (2020) Smart home energy management system—a review. Adv Build Energy Res (in press)
28. Delfino F, Ferro G, Minciardi R, Robba M, Rossi M, Rossi M (2019) Identification and optimal control of an electrical storage system for microgrids with renewables. Sustain Energy Grids Netw 17:100183

29. Hooshmand R, Bornapour M, Khodabakhshian A (2019) A comprehensive stochastic energy management system of micro-CHP units, renewable energy sources and storage systems in microgrids considering demand response programs. Renew Sustain Energy Rev 108:355–368
30. Delfino F, Ferro G, Robba M, Rossi M (2019) An energy management platform for the optimal control of active and reactive power in sustainable microgrids. IEEE Trans Ind Appl 55(6):7146–7156
31. Jianhui WS, Wang JR (2014) Stochastic energy scheduling in micro-grids with intermittent renewable energy resources. IEEE Trans Smart Grid 5:1876–1883
32. You P, Yang Z, Chow MY, Sun Y (2015) Optimal cooperative charging strategy for a smart charging station of electric vehicles. IEEE Trans Power Syst 1–11
33. Ferro G, Laureri F, Minciardi R, Robba M (2018) An optimization model for electrical vehicles scheduling in a smart grid. Sustain Energy Grids Netw 14:62–70
34. Ferro G, Laureri F, Minciardi R, Robba M (2019) A predictive discrete event approach for the optimal charging of electric vehicles in microgrids. Control Eng Pract 86(2018):11–23
35. Ferro G, Minciardi R, Parodi L, Robba M (2020) Discrete event optimization of a vehicle charging station with multiple sockets. Discrete Event Dyn Syst Theory Appl

Chapter 3
Optimal Charging in Smart Grids: Discrete-Time Optimization Model for Aperiodic Scheduling

3.1 Introduction

In this chapter, attention is focused on formalizing a discrete-time optimization problem for the scheduling of electric vehicles. Specifically, all variables and parameters are discretized in time (over a fixed time interval of length T).

The choice of the time interval length and the time horizon is particularly important because it influences the number of decision variables and constraints and thus has implications on the run time of the optimization problem solution. Optimization problems for the scheduling of electric vehicles are generally inserted within the energy management systems (EMSs) and modeled in discrete time [1–5]. EVs are typically considered as forecasted static loads in the decision models without considering the decisions related to their scheduling based on arrival, desired energy, and due dates. The reason is that EVs are (at least nowadays) few over the territory. Often a reservation service is not available, and EMSs do not exploit the similarities that the charging process has with the scheduling of manufacturing systems. In a future in which EVs will increase considerably, there will be a need for designing and managing the process in a more structured way to limit queues and negative impacts on the electrical grid. Such an approach is particularly interesting in the case of demand response programs, i.e., when there is the necessity of shifting power demands to help the distribution grid manager and/or receive remuneration from a market aggregator.

In the literature, energy management systems (EMSs) based on optimization models are used for the operational management of production and storage systems and also to manage demand response programs (see, for example, the works proposed in [6, 7]). EVs are generally considered as additional loads, and their schedule is not detailed. Usually, optimization-based EMSs are used to define the optimal operational management including cost and environmental performance indexes for a set

G. Ferro et al., *Optimization of Electric-Vehicle Charging*, Advances in Industrial Control, https://doi.org/10.1007/978-3-031-61917-5_3

of different technologies (renewables, EVs, traditional sources, and storages). In this connection, Zhao et al. [8] review different charging methodologies and strategies. Instead, the work of Deilami et al. [9] investigates the contribution of plug-in EVs in smart grids. Allard et al. [10] compare the impact of a smart charging strategy to traditional EVs' charging in power grids with large penetration of renewables. The authors in [11] develop a strategy to regulate the vehicle's charging process, introducing different priorities for the various objectives, like achieving maximum flexibility, the effective management of the power flows to smooth the demand curve, and increasing the system's power quality.

The study presented in [12] represents the starting point for the model presented in the following. In particular, in this chapter, the integration of EVs in smart grids is addressed. The decision problem is here characterized by the definition of the optimal charging strategies to feed electric vehicles and the scheduling of production plants and storage systems. Specifically, the charging station is considered as a machine that has to schedule different jobs (i.e., EVs) whose sequence is fixed (according to the arrival order). Still, the charging pattern and time have to be defined. The approach is inspired by approaches usually adopted in smart manufacturing systems. The decision variables are represented by optimal scheduling of power plants and the recharging process of EVs while different classes of constraints have been introduced.

The resulting optimization problem is deterministic, nonlinear, and includes continuous and binary variables, and it is based on forecasts as regards demands and renewable power availability.

A control scheme using a receding horizon approach (refer to Fig. 3.1) has been implemented to address the optimization problem and gather real-time data from the field for forecasting updates. The selection of both H (optimization) and S (simulation) horizons is dependent on the information available on uncertain forecasts/parameters such as current states, predicted demands, and prices. The length of the optimization horizon determines the running time of the optimization problem. The receding horizon approach retains solely the solution relating to the initial time interval during the first run, followed by the acquisition of new information (new data from field sensors and forecasts), and then by the run of a new optimization problem. This approach allows for the continuous acquisition of new information as it becomes available.

A microgrid's system model is presented in Sect. 3.2. Then, Sect. 3.3 reports the modeling of EVs. Finally, in Sect. 3.4, the optimization problem is formalized and then applied to a case study (Sect. 3.5).

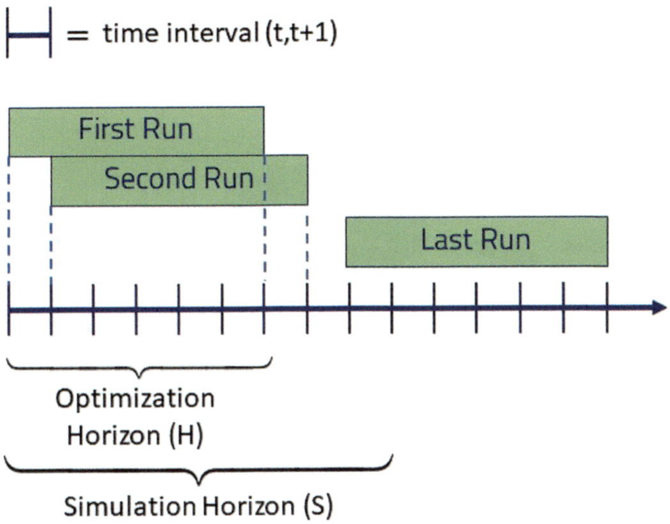

Fig. 3.1 The receding horizon control scheme

3.2 The System Model[1]

The considered system consists of a local area composed of the following elements (see Fig. 3.2): production from renewable resources (wind turbines and photovoltaics), internal combustion engines, and electrical storage systems; a connection with the main grid, a charging station (CS) for EVs, and electrical loads.

The forecasts in each time interval (for all the duration of the optimization horizon) of power produced by renewable resources have been calculated based on meteorological forecasting and/or black-box models and machine learning techniques. To define the mathematical model, the following sets are introduced:

- $F = \{1, \ldots, L\}$ set of fossil fuel generators;
- $S = \{1, \ldots, M\}$ set of storage systems;
- $\mathrm{EV} = \{1, \ldots, N\}$ set of electric vehicles.

For photovoltaics, it is necessary to predict solar irradiation $G(t_k)$ in a time interval (t_k, t_{k+1}) and then to calculate the power production $P_{\mathrm{PV}}(t_k)$ based on the solar flat plate collector area A and the efficiency of the solar flat plate collector η_{PV}. That is,

$$P_{\mathrm{PV}}(t_k) = \eta_{\mathrm{PV}} A_{\mathrm{PV}} G(t_k) \quad k = 0, \ldots, T-1. \tag{3.1}$$

For wind turbines, it is necessary to predict weed speed at a specific site. Then, $[v_c, v_r]$ being an interval of velocities, the following wind model is used to calculate power output $P_W(t_k)$ as a function of the current wind speed at the hub height $v(t_k)$:

[1] Text in Sect. 3.2 is © 2018 Elsevier Ltd., reproduced from [12], https://doi.org/10.1016/j.segan.2018.04.002; used with permission.

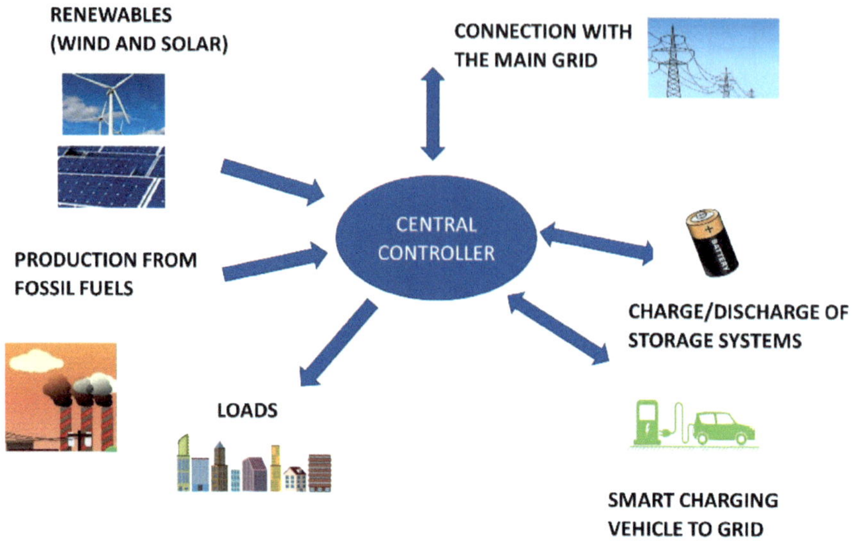

Fig. 3.2 The considered system

$$P_W(t_k) = \begin{cases} 0 & v(t_k) < v_c \\ P_r(a + bv(t_k)) & v_c \leq v(t_k) \leq v_r \\ P_r & v_r \leq v(t_k) \leq v_f \\ 0 & v(t_k) > v_f \end{cases} \quad k = 0, \ldots, T-1. \tag{3.2}$$

It should be observed that wind speed v_t is considered at the wind turbine hub height H_{hub} and not at the height H_{data} of which data ($v_{\text{data}}(t_k)$) are available. Thus, taking into account the surface roughness length z_0 [m], velocity can be derived for example by $v(t_k) = v_{\text{data}}(t_k)\frac{\ln(H_{\text{hub}}/z_0)}{\ln(H_{\text{data}}/z_0)}$. The power balance constraint that should be satisfied is given by

$$P_W(t_k) + P_{\text{PV}}(t_k) + P_{\text{grid}}(t_k) + \sum_{l=1}^{L} P_{F,l}(t_k) + \sum_{m=1}^{M} P_{S,m}(t_k) = P_D(t_k)$$
$$+ \sum_{i=1}^{N} P_{\text{EV},i}(t_k) \quad k = 0, \ldots, T-1 \tag{3.3}$$

where $P_{\text{grid}}(t_k)$ is power exchanged with the main grid, $P_{F,l}(t_k)$ is the power from the lth fossil fuel production plant, $P_{S,m}(t_k)$ is the power exchange with the mth storage system, $P_{\text{EV},i}(t_k)$ is the power exchange with the ith electric vehicle connected to the charging station, and $P_D(t_k)$ is the power demand.

Each storage system can be represented by the following state equation (active sign convention is considered):

$$x_m(t_{k+1}) = x_m(t_k) - \eta_{m,\text{out}} \frac{P_{S,m,\text{out}}(t_k)}{\text{CAP}_m} \Delta t + \eta_{m,\text{in}} \frac{P_{S,m,\text{in}}(t_k)}{\text{CAP}_m} \Delta t$$
$$k = 0, \ldots, T-1;\ m = 1, \ldots, M \tag{3.4}$$

where Δt is time interval length, $x_m(t_k)$ is the state-of-charge of the mth battery with a capacity CAP_m [kWh], $\eta_{m,\text{out}}$ is efficiency in discharging mode, and $\eta_{m,in}$ is efficiency in charging mode.

Note that $P_{S,m,\text{out}}(t_k)$ is power exchange during discharging and $P_{S,m,\text{in}}(t_k)$ is power exchange during charging, being

$$P_{S,m,\text{out}}(t_k) - P_{S,m,\text{in}}(t_k) = P_{S,m}(t_k) \quad k = 0, \ldots, T-1;\ m = 1, \ldots, M. \tag{3.5}$$

The following disjunctive inequalities are introduced to keep the mutual exclusiveness of $P_{S,m,\text{out}}(t_k)$ and $P_{S,m,\text{in}}(t_k)$ terms:

$$P_{S,m,\text{out}}(t_k) \leq \theta_m(t_k) M \quad k = 0, \ldots, T-1;\ m = 1, \ldots, M \tag{3.6}$$

$$P_{S,m,\text{in}}(t_k) \leq (1 - \theta_m(t_k)) M \quad k = 0, \ldots, T-1;\ m = 1, \ldots, M \tag{3.7}$$

where M is a big number and $\theta_m(t_k)$ is an auxiliary binary variable (equal to one if the battery is discharging and 0 if it is discharging).

3.2.1 Piecewise Linear Storage Modeling

In real storage systems, the maximum power level depends on the state-of-charge. Specifically, we consider a storage system characterized by its battery management system that receives commands from a central controller like an EMS for a microgrid. Then, the maximum power that can be requested is lower than the rated value, for certain values of the state-of-charge.

This phenomenon can be expressed through a piecewise linear formulation as follows:

$$P_{S,b,m}(x_m(t_k)) = \begin{cases} \mu P_{S,m}^{\text{sp}} & \text{if } x_m(t_k) \leq a_m \\ c_m x_m(t_k) + d_m & \text{if } x_m(t_k) \geq a_m \end{cases} \quad k = 0, \ldots, T-1;\ m = 1, \ldots, M \tag{3.8}$$

where $P_{S,m}^{\text{sp}}$ is the power set point [kW], and a_m, c_m, d_m are parameters typical of the specific storage system. The model (3.8) allows defining a polygon (see Fig. 3.3) that includes the values of the power and the state-of-charge (SOC) that are acceptable (the light blue area in the figure).

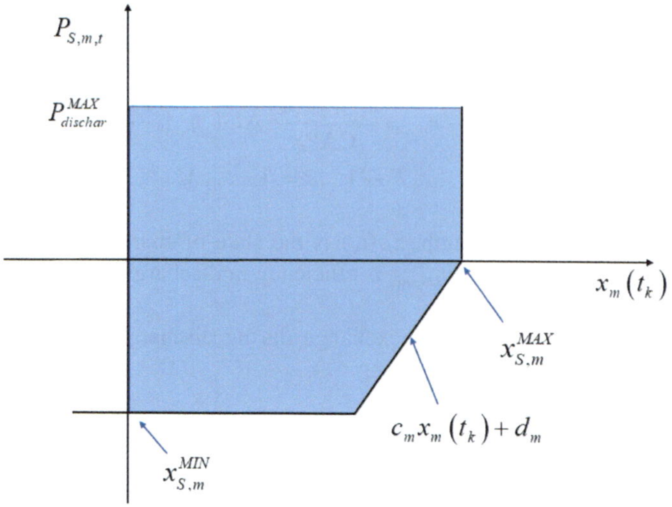

Fig. 3.3 Piecewise storage constraints

The feasible area represented in Fig. 3.3 is defined by the constraints

$$P_{S,b,m}(x_m(t_k)) \le P_{S,m}(t_k) \le P_{\text{dischar}}^{\text{MAX}} \quad k = 0, \ldots, T-1; \; m = 1, \ldots, M \tag{3.9}$$

$$x_{S,m}^{\text{MIN}} \le x_m(t_k) \le x_{S,m}^{\text{MAX}} \quad k = 0, \ldots, T-1; \; m = 1, \ldots, M \tag{3.10}$$

where $x_{S,m}^{\text{MIN}}$ and $x_{S,m}^{\text{MAX}}$ are the minimum and maximum working limits for the state-of-charge, $P_{\text{dischar}}^{\text{MAX}}$ is the maximum power the storage system can deliver when discharging (assumed constant), and $P_{S,b,m}(x_m(t_k))$ is the (negative) lower limit of the power the storage can absorb when recharging, expressed as a function of the state-of-charge. Thus, in the optimization problem, it is necessary to add the following constraints:

$$0 \le P_{S,m,\text{out}}(t_k) \le P_{\text{dischar}}^{\text{MAX}} \quad k = 0, \ldots, T-1; \; m = 1, \ldots, M \tag{3.11}$$

$$0 \le P_{S,m,in}(t_k) \le -P_{S,b,m}(x_m(t_k)) \quad k = 0, \ldots, T-1; \; m = 1, \ldots, M. \tag{3.12}$$

3.3 Modeling Electric Vehicles Scheduling[2,3]

The EVs are identified by index i, which is assigned according to the arrival order. Since we are using a discrete-time approach, all variables are considered in discrete-time instants t_k, with a unit length of Δt. Finally, the overall optimization horizon is (t_0, t_T) and consists of T time intervals.

The charging process is defined by two different time intervals:

- $(t_{C_i}, t_{C_{i+1}})$, i.e., the starting interval in which EV i begins the charging procedure;
- $(t_{f_i}, t_{f_{i+1}})$, i.e., the final interval in which EV i concludes the charging procedure.

The following data characterize each EV:

- t_{rel_i} the release time in which the vehicle is available to charge;
- $t_{\text{dd}_{i+1}}$ the due-date time at which the charging process of the vehicle *should* be concluded;
- $t_{\text{dl}_{i+1}}$ the deadline time at which the charging process of the vehicle *must* be concluded;
- α_i the tardiness penalty cost [€/h].

The energy [kWh] stored in each vehicle is governed by first-order equation:

$$x_i(t_{k+1}) = x_i(t_k) - \eta_{i,\text{out}} \frac{P_{\text{EV,out},i}(t_k)}{\text{CAP}_i} \Delta t + \eta_{i,\text{in}} \frac{P_{\text{EV,in},i}(t_k)}{\text{CAP}_i} \Delta t \quad k = 0, \ldots, T-1; \quad i = 1, \ldots, N \tag{3.13}$$

$$P_{\text{EV,in},i}(x_i(t_k)) = \begin{cases} \mu P^{\text{sp}}_{\text{EV},i} & \text{if } x_i(t_k) \le a_{v,i} \\ c_{v,i} x_i(t_k) + d_{v,i} & \text{if } x_i(t_k) \ge a_{v,i} \end{cases} \quad k = 0, \ldots, T-1; \; i = 1, \ldots, N \tag{3.14}$$

$$0 \le P_{EV,\text{in},t} \le -P_{EV,i}(x_m(t_k)) \quad k = 0, \ldots, T-1; \; i = 1, \ldots, N \tag{3.15}$$

$$0 \le P_{\text{EV,out},i}(t_k) \le P^{\text{MAX}}_{\text{EV,dischar},i} \quad k = 0, \ldots, T-1; \; i = 1, \ldots, N \tag{3.16}$$

where $P_{\text{EV,out},i}(t_k)$ is power exchange during discharging and $P_{\text{EV,in},i}(t_k)$ is power exchange during charging, being

$$-P_{\text{EV,out},i}(t_k) + P_{\text{EV,in},i}(t_k) = P_{\text{EV},i}(t_k) \quad k = 0, \ldots, T-1; \quad i = 1, \ldots, N. \tag{3.17}$$

[2] Text in Sect. 3.3 is © 2018 Elsevier Ltd., reproduced from [12], https://doi.org/10.1016/j.segan.2018.04.002; used with permission.

[3] Text in Sect. 3.3 is © Federica Laureri, reproduced from [13], https://dx.doi.org/10.15167/laureri-federica_phd2018-05-30; used with permission.

The following disjunctive inequalities are introduced to keep the mutual exclusiveness of $P_{EV,in,i}(t_k)$ and $P_{EV,out,i}(t_k)$ terms:

$$P_{\mathrm{EV,out},i}(t_k) \leq \psi_i(t_k)M \quad k = 0, \ldots, T-1; \quad i = 1, \ldots, N \tag{3.18}$$

$$P_{\mathrm{EV,in},i}(t_k) \leq (1 - \psi_i(t_k))M \quad k = 0, \ldots, T-1; \; i = 1, \ldots, N \tag{3.19}$$

where M is a big number, and $\psi_i(t_k)$ is an auxiliary binary variable (equal to one if the battery is discharging and 0 if it is discharging).

3.3.1 The Overall Optimization Problem

The objective function includes four different terms of cost:

- C_{eng}: non-renewable sources operating cost;
- C_{purchase}: main grid energy purchase cost;
- C_{sell}: main grid energy selling benefit;
- C_{tard}: tardiness cost for each EV.

The control variables for the optimization problem are

- $P_{\mathrm{grid},t}$ (unrestricted in sign): the power exchanged with the main grid [kW];
- $P_{F,l}(t_k)$: the power from the lth fossil fuel production plant;
- $P_{S,m}(t_k)$ (unrestricted in sign): the power exchange with the mth storage system;
- $P_{\mathrm{EV},i}(t_k)$ (unrestricted in sign): the power exchange with the ith electric vehicle connected to the charging station;
- c_i, f_i: time in which vehicle i starts (finishes) the charging process;
- $\delta_i(t_k)$: binary variable equal to 1 if vehicle i is connected to the charging station in a time interval (t_k, t_{k+1}) and 0 otherwise.

The state variables for the optimization problem are

- $x_i(t_k)$: state-of-charge of ith vehicle at time t_k;
- $x_i(t_k)$: state-of-charge of ith vehicle at time t_k.

The overall objective function is given by

$$\min \; (C_{\mathrm{eng}} + C_{\mathrm{purchase}} - C_{\mathrm{sell}} + C_{\mathrm{tard}}) \tag{3.20}$$

where

- $C_{\mathrm{eng}} = \sum_{k} \sum_{l} \gamma P_{F,l}(t_k)\Delta t$ are the energy overall production costs for all non-renewables, where γ is the unit cost;

- $C_{\text{purchase}} = \sum_{k=0}^{T-1} \beta_p(t_k) \max(P_{\text{grid}}(t_k), 0)\Delta t$ is the cost paid to buy energy from the main grid, with $\beta_p(t_k)$ the time-varying unit price;
- $C_{\text{sell}} = \sum_{k=0}^{T-1} \beta_s(t_k) \max(-P_{\text{grid}}(t_k), 0)\Delta t$ is the income due to the sale of energy to the main grid, with $\beta_s(t_k)$ the time-varying unit benefit;
- $C_{\text{tard}} = \sum_{i=1}^{N} \alpha_i \max(t_{f,i} - t_{\text{dd},i}, 0)$ is the total tardiness cost with respect to the due date.

The optimization problem is bounded by the feasibility region given by Eqs. (3.1)–(3.19), and by equations that are necessary to satisfy technical issues.

First, it is necessary to avoid that more than one vehicle is charging:

$$\sum_{i=1}^{N} \delta_i(t_k) \leq 1 \quad k = 0, \dots, T-1. \tag{3.21}$$

Let.

- (c_i, c_{i+1}) is the start charging interval of EV i;
- (f_i, f_{i+1}) is the final charging interval of EV i.

We need to introduce a constraint that imposes the consistency of the beginning of the charging process with the release time

$$c_i \geq t_{\text{rel},i} \quad i = 1, \dots, N. \tag{3.22}$$

Similarly, the charging procedure must end before the deadline

$$f_i \leq t_{\text{dl}_i} + 1 \quad i = 1, \dots, N. \tag{3.23}$$

A further constraint has the function of relating the state of the charging station with the time interval during which the charging of a vehicle i takes place:

$$\begin{aligned}
&\text{if } (t_{\text{rel},i} - t_k) \geq 0 \quad \rightarrow \quad \delta_i(t_k) = 0 \\
&\text{if } (t_{\text{rel},i} - t_k) \geq 0 \quad \text{or} \quad (t_k - (f_i + 1)) \geq 0 \quad \rightarrow \quad \delta_i(t_k) = 0 \\
&\text{if } (t_{\text{rel},i} - t_k) \leq 0 \quad \text{and} \quad (t_k - (f_i + 1)) \leq 0 \quad \rightarrow \quad \delta_i(t_k) \in [0, 1] \\
&\qquad k = 0, \dots, T-1; \; i = 1, \dots, N.
\end{aligned} \tag{3.24}$$

Constraints (3.24) can be substituted by the following linear constraints:

$$(1 - \delta_i(t_k))M \geq (t_{\mathrm{rel},i} - t_k) \quad k = 0, \ldots, T-1; \quad i = 1, \ldots, N \tag{3.25}$$

$$(1 - \delta_i(t_k))M \geq (t_k - f_i - 1) \quad k = 0, \ldots, T-1; \; i = 1, \ldots, N. \tag{3.26}$$

Afterwards, we impose that the EV leaves the charging station only when reaching the desired state-of-charge:

$$x_{i,f_i+1} = \mathrm{ER}_i \quad k = 1, \ldots, T; \quad i = 1, \ldots, N. \tag{3.27}$$

At the release time, the state-of-charge must be equal to the initial one:

$$x_i\left(t_{\mathrm{rel}_i}\right) = x_{i,\mathrm{in}} \quad k = 1, \ldots, T; \quad i = 1, \ldots, N. \tag{3.28}$$

Moreover, the power exchange with the external should be limited within the minimum and maximum values:

$$-P_{\mathrm{grid}}^{\mathrm{MAX}} \leq P_{\mathrm{grid}}(t_k) \leq P_{\mathrm{grid}}^{\mathrm{MAX}} \quad k = 0, \ldots, T-1. \tag{3.29}$$

Similarly, a maximum power constraint must be considered for production plants from fossil fuels. That is,

$$0 \leq P_{F,l}(t_k) \leq P_F^{\mathrm{MAX}} \quad k = 0, \ldots, T-1; \quad l = 1, \ldots, L. \tag{3.30}$$

Finally, the following inequalities are introduced to link the binary variable $\delta_i(t_k)$ with the EV's power limits:

$$0 \leq P_{\mathrm{EV,out},i}(t_k) \leq \delta_i(t_k) P_{\mathrm{EV,dischar},i}^{\mathrm{MAX}} \quad k = 0, \ldots, T-1; \quad i = 1, \ldots, N \tag{3.31}$$

$$P_{\mathrm{EV,in},i}(t_k) - M\,\delta_i(t_k) \leq 0 \quad k = 0, \ldots, T-1; \quad i = 1, \ldots, N. \tag{3.32}$$

3.4 Case Study Application

In this section, the developed optimization problem is applied to a case study. There are two traditional plants, one storage system, one photovoltaic plant, one wind turbine, one charging station, and electrical demand.

Data are reported in Table 3.1 (parameters of the microgrid) and Fig. 3.4 (forecasted power demand and production from photovoltaics) for a whole day.

Table 3.1 Microgrid data

Parameter	Value	Parameter	Value
$P_{F,1}^{\max}$	30 [kW]	c_S	0.6
$P_{F,2}^{\max}$	65 [kW]	d_S	80
CAP	100 [kWh]	$x_{S,m}^{\mathrm{MIN}}$	0.2
$\eta_{S,\mathrm{out}}$	1.1	$x_{S,m}^{\mathrm{MAX}}$	0.8
$\eta_{S,in}$	0.9	$P_S^{\mathrm{sp}} = P_{\mathrm{dischar}}^{\mathrm{MAX}}$	35 [kW]
M	1000	γ	0.2 [€/kWh]
μ_S	1	$P_{\mathrm{grid}}^{\mathrm{MAX}}$	300 [kW]

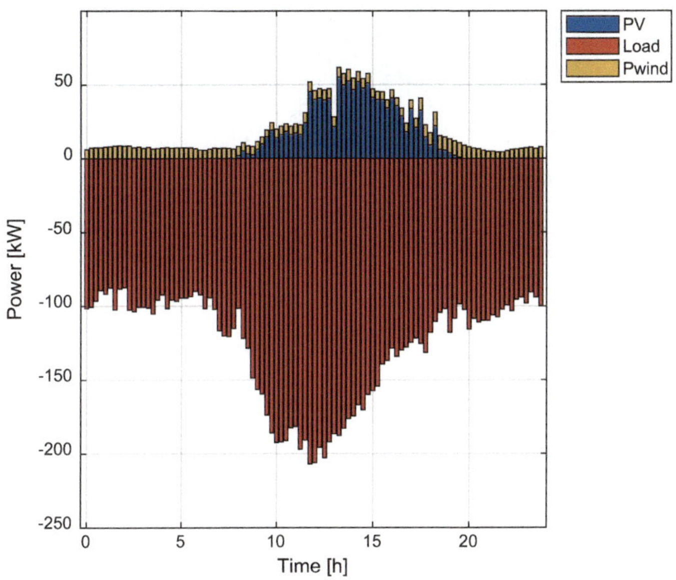

Fig. 3.4 Load and renewable generation

The piecewise constant pattern of unit cost $\beta_p(t_k)$ [€/kWh] for the energy purchased by the external grid is shown in Fig. 3.5, while the energy selling price $\beta_s(t_k)$ is kept constant at 0.08 [€/kWh].

The optimization problem (3.3)–(3.32) has been implemented by using commercial optimization LINGO. The overall optimization horizon consists of 24 h and a time interval of 15 min ($\Delta t = 0.25$ [h]). In the proposed simulation, four vehicles are considered, whose data are reported in Table 3.2.

This particular instance of the scheduling problem has been solved in 33 [s] with a cost of 540.12 [€]. The detailed scheduling of the microgrid is reported in Fig. 3.6.

It is important to note that the power into the storage system and the vehicles is negative since active sign convention is used. Figure 3.6 shows a consistent behavior

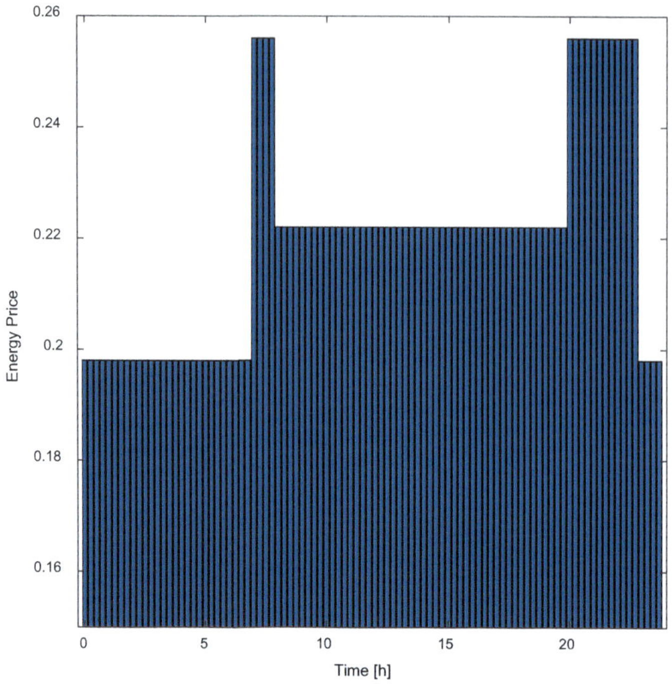

Fig. 3.5 Buying price pattern

Table 3.2 EVs' data

Parameter	Vehicle 1	Vehicle 2	Vehicle 3	Vehicle 4
$x_{i,\text{in}}$	0	0	0	0
ER_i	0.6	0.75	0.95	0.53
CAP_i	22 [kWh]	25 [kWh]	31 [kWh]	22 [kWh]
$P_{\text{EV,dischar},i}^{\text{MAX}}$	13 [kW]	13 [kW]	13 [kW]	13 [kW]
$t_{\text{rel},i}$	1.25 [h]	2.5 [h]	4.5 [h]	8.75 [h]
t_{dd_i}	5.5 [h]	9 [h]	12.5 [h]	15 [h]
t_{dl_i}	7.5 [h]	12 [h]	15.75 [h]	17.5 [h]
α_i	0.1 [€/h]	0.1 [€/h]	0.1 [€/h]	0.1 [€/h]

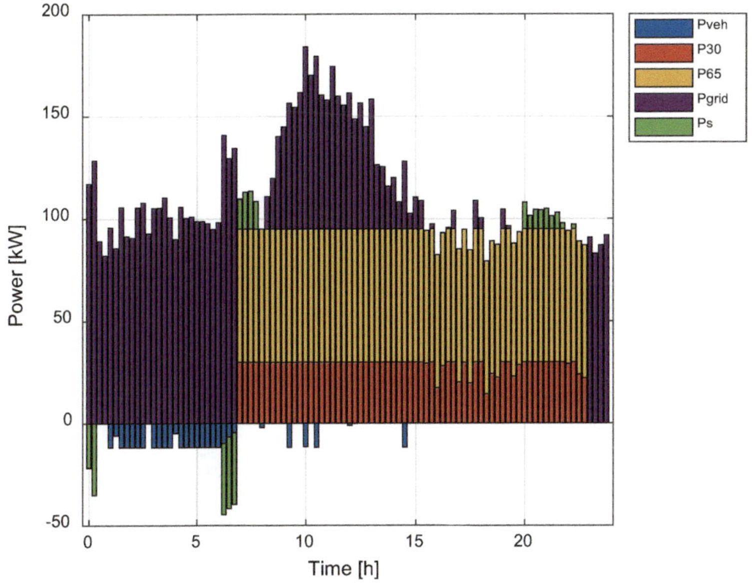

Fig. 3.6 Optimal microgrid scheduling

of the traditional generators concerning the buying price pattern, as a matter of fact, is used only when the fuel price γ is less than $\beta_p(t_k)$.

As regards the charging vehicles, scheduling is reported in Fig. 3.7. Each vehicle is recharged with respect to the timing constraints. In particular, the charging power profile of vehicle 3, clearly shows the activation of the constraints related to the battery charge (3.14): when the SOC is close to 90%, the power absorbed is far from the maximum (i.e., 13 [kW]).

Finally, Fig. 3.8 depicts the Gantt diagram of the charging station (corresponding to the evolution of $\delta_i(t_k)$ variable). As can be seen from this figure, the process of vehicle 3 is preempted with respect to vehicle 4.

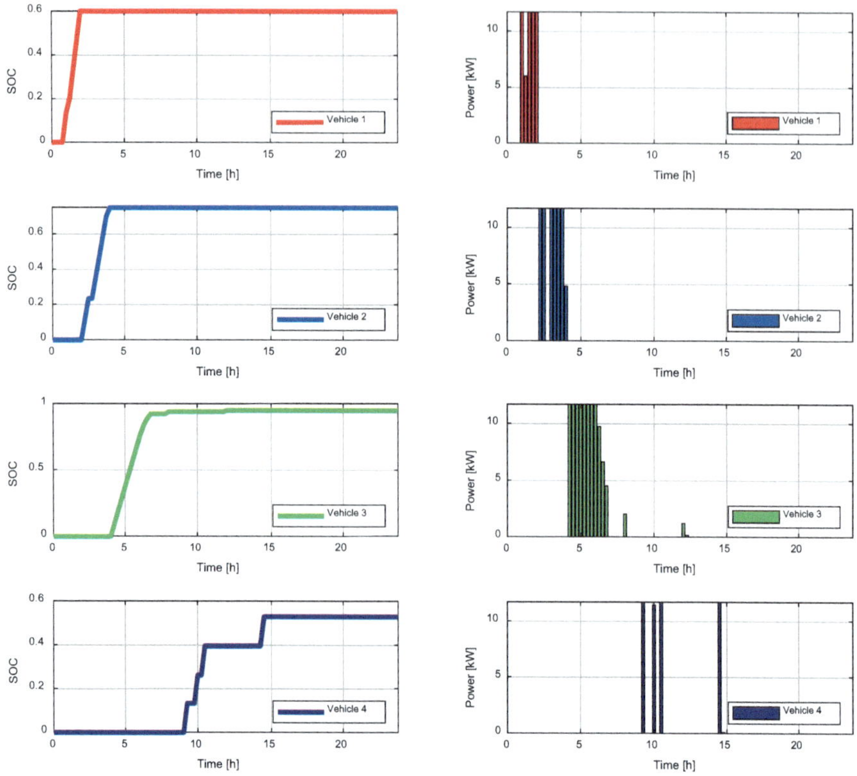

Fig. 3.7 Optimal EV scheduling

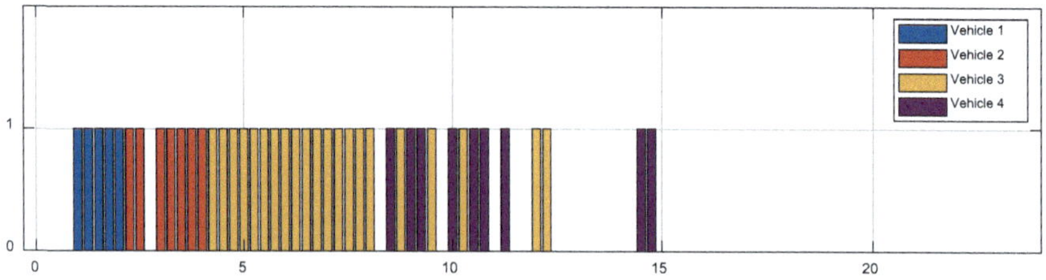

Fig. 3.8 Charging station Gantt diagram

References

1. Parisio A, Rikos E, Glielmo L (2014) A model predictive control approach to microgrid operation optimization. IEEE Trans Control Syst Technol 22(5):1813–1827
2. Minciardi R, Robba M (2017) A bi-level approach for the stochastic optimal operation of interconnected microgrids. IEEE Trans Autom Sci Eng 14(2):482–493
3. Bracco S, Delfino F, Pampararo F, Robba M, Rossi M (2015) A dynamic optimization-based architecture for polygeneration microgrids with tri-generation, renewables, storage systems and electrical vehicles. Energy Convers Manag 96:511–520
4. Zia MF, Elbouchikhi E, Benbouzid M (2018) Microgrids energy management systems: a critical review on methods, solutions, and prospects. Appl Energy 222:1033–1055
5. Zhang, Gatsis N, Giannakis G (2013) Robust energy management for microgrids with high-penetration renewables. IEEE Trans Sustain Energy 4(4):944–953
6. Delfino F, Ferro G, Minciardi R, Robba M, Rossi M, Rossi M (2019) Identification and optimal control of an electrical storage system for microgrids with renewables. Sustain Energy Grids Netw 17:100183
7. Delfino F, Ferro G, Robba M, Rossi M (2019) An energy management platform for the optimal control of active and reactive powers in sustainable microgrids. IEEE Trans Ind Appl 55(6):7146–7156
8. Zhao, Awater, Schafer, Breuer, Moser (2011) Scenario-based evaluation on the impacts of electric vehicle on the municipal energy supply systems. In: IEEE Power and Energy Society general meeting, pp 1–8
9. Deilami M, Moses M (2011) Real-time coordination of plug-in electric vehicle charging in smart grids to minimize power losses and improve voltage profile. IEEE Trans Smart Grid 2(3):456–467
10. Allard S, See PC, Molinas M, Fosso OB, Foosnas JA (2013) Electric vehicles charging in a smart microgrid supplied with wind energy. In: IEEE PowerTech (POWERTECH), Grenoble, pp 1–5
11. Khemakhem S, Rekik M, Krichen L (2017) A flexible control strategy of plug-in electric vehicles operating in seven modes for smoothing load power curves in smart grid. Energy 118:197–208
12. Ferro G, Laureri F, Minciardi R, Robba M (2018) An optimization model for electrical vehicles scheduling in a smart grid. Sustain Energy Grids Netw 14:62–70
13. Laureri F. Methods for optimal microgrid management. PhD Thesis [Online]. Available at: https://doi.org/10.15167/laureri-federica_phd2018-05-30

Chapter 4
Optimal Charging in Smart Grids: A Discrete-Event Approach for Aperiodic Scheduling in Single Socket Charging Stations

4.1 Introduction: Discrete-Event Systems

Several systems in very different application areas deal with discrete variables (typically assuming integer values) and with processes that depend on instantaneous "events" such as the pushing of a button, the phase transition of a traffic light, and the on/off transition of a portion of the grid or of a charging station.

A discrete-event system is characterized by discrete-valued state variables, and by a set of classes of events that can occur and drive the system behavior. To each class of event, a state transition is associated, determining the change in the values of the state variables. An event can be identified by a specific action from the external or some spontaneous occurrence determined by the system's nature (the completion of a service process, the occurrence of a failure, etc.).

In discrete-state systems, the state changes only at certain time instants through instantaneous transitions. The timing mechanism based on which the event takes place can be

- *Time-driven systems*: state transitions are supposed to take place only at time instants known in advance (typically, discrete-valued and separated by equal time intervals). If no event occurs during the last time interval, the state does not change, and the process proceeds until the next time instant (i.e., the end of the successive time interval). State transitions are, in this case, synchronized.
- *Event-driven systems*: every event has its own occurrence time instant, and state transitions are immediately implemented. The resulting process is asynchronous.

A discrete-event system (DES) is a discrete-state, event-driven system in which the evolution of its state entirely depends on the occurrence of asynchronous discrete events over time. Thus, the state can only change at discrete points in time. As a consequence, time is not an independent variable.

G. Ferro et al., *Optimization of Electric-Vehicle Charging*, Advances in Industrial Control, https://doi.org/10.1007/978-3-031-61917-5_4

The continuous-time state equations are no longer valid, and the state equations for DES should be determined.

Different systems can be represented like DES, such as queueing, computers, communication, traffic, and manufacturing. Often, a DES representation can help in reducing the number of variables in optimization problems and thus improve the computational time. In recent literature, Miao [1] focuses on prioritized DESs with real-time constraints, motivated by applications in power-limited systems where a trade-off exists between resource efficiency and system performance. For off-line control, structural properties of the optimal sample path have been found to reduce the search space of the optimal task execution sequence. In contrast, a receding horizon approach has been adopted for online control. In [2], a methodology is proposed to deal with planning problems in flexible manufacturing systems for planning in large batches of production. The proposed methodology was tested in a plant of moderate size. The results show that planning for batches as large as desired can be achieved efficiently at a very reduced computational cost. In [3], a discrete-event approach controls temperature through HVAC in buildings. An event-based control adjusts actions when certain events occur, which may be faster and more scalable than state-based or time-driven control methods. Since the choice of events is a trade-off between the computational efficiency and the control performance, the authors study events for the HVAC control problem and define how to select events that capture sufficient state information with relatively small event space.

In the next subsections, a discrete-event approach is proposed for the optimal scheduling of electric vehicles in which vehicles are the customers and charging stations are the machines. The resulting optimization model is applied to a case study. Specifically, the rest of the chapter is organized as follows. Section 4.2 describes the application of DES to EVs. Section 4.3 describes the system, while Sects. 4.4 and 4.5 report the objective function and constraints. Finally, the application to a case study is described.

4.2 Discrete-Event Approaches for Electric Vehicles[1]

The operational management of a grid with production plants and storage systems to satisfy an electrical demand is a problem usually faced through discrete-time models, with an optimization horizon and a time interval. However, a discrete-time formalization could lead to a high computational effort if the target is to schedule electric vehicles.

In this chapter, a discrete-event approach based on the works presented in [4, 5] is considered in the operational management of a microgrid. The electrical grid architecture comprehends a single node where all the loads and generation units are connected. The objective function is composed of three terms:

---

[1] The text in this section is © 2019 Elsevier. Based on [5], https://doi.org/10.1016/j.conengprac.2019.02.004; reproduced by permission.

- the cost due to the energy production from non-renewable sources;
- the cost/gain for energy bought/sold from/to the main grid;
- the penalty cost for tardiness.

The system behavior is described as a discrete sequence of state transitions that coincides with discrete events. Thus, the system state immediately changes when an event occurs. The discrete-event approach has been chosen because it reduces the number of variables in the optimization problem. The authors in [6] confirm that a discrete-time approach implies many decision variables when time intervals grow in number. Conversely, a discrete-event approach can significantly reduce the number of variables and track the system behavior whenever an event takes place [4, 5].

However, in microgrid management, the discrete-event approach has some difficulties, mainly related to the forecasted powers which are usually given as discrete-time quantities. This makes mandatory the introduction of some hypotheses.

Certainly, the problem here considered and formalized falls within the broad class of scheduling problems [7–9]. A scheduling problem generally includes the assignment, sequencing, and timing of a given set of jobs to a given set of resources, to provide a certain service for these jobs. In general, the solution of a scheduling problem requires the determination of optimal decisions as regards assignment, sequencing, and timing. In this chapter, only the timing decisions are considered.

In the literature, it is common to find event-driven approaches for the management of EVs in the microgrid framework. Some examples are given by [10–12] where many different approaches for the optimal charging of EVs are investigated. However, these approaches are online scheduling algorithms which can correct a previously determined schedule when something (an event) happens making this correction necessary. These events could correspond to the arrival of a new EV that necessitates being charged, or the occurrence of some failure in a charging station.

Instead, the approach described in this chapter aims to optimize the charging schedule of a given set of EVs.

The main features of the optimization model are the following:

- the optimization objective is that of finding the optimal schedule for the charging process of a set of EVs in a grid-connected microgrid;
- the schedule optimization problem is formalized modeling the dynamics of the system within a discrete-event setting;
- the approach followed is based on the predictive control paradigm; that is, the external (non-manipulable) inputs are taken as deterministic inputs, on the basis of their forecasts available at the current time; then, at each control step, the same optimization problem, on a time horizon of constant length (shifted forward at each step), is stated and solved, making use of updated input forecasts; in the considered model, the non-manipulable inputs correspond to power coming from renewables and external local loads.

The proposed approach is then tested through the application on a real case study that corresponds to a complex infrastructure, located in the Savona Campus (University of Genoa), that includes the connection to the main grid, different renewable

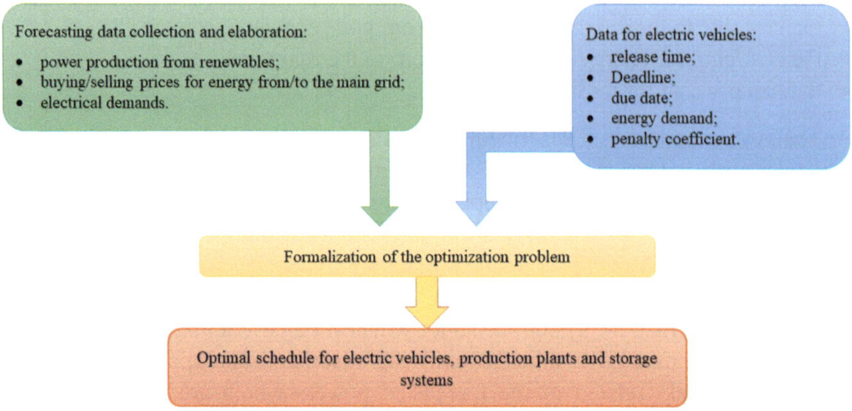

Fig. 4.1 Representation of the overall schedule optimization procedure

source plants, an electrical storage system, a conventional plant for power generation, and a vehicle charging station.

Figure 4.1 presents the scheme of the proposed approach. The procedure to obtain the schedule of the charging of the vehicles can be viewed as composed of the following steps:

- information is collected in discrete time as regards: forecasts for production from renewables, loads, and buying/selling energy prices; the initial state-of-charge for storage systems is observed;
- analytical interpolation of data regarding forecasting (buying/selling energy prices, renewables, loads);
- data collection for electric vehicles: release times, deadlines, due dates, sequence of arrivals, energy requests, and penalties for tardiness;
- formalization and implementation of the optimization problem;
- determination of the optimal schedule for conventional power generation plants, storage systems, and charging of electric vehicles.

As a matter of fact, it is apparent that the problem concerning optimal scheduling of charging of EVs falls within the class of scheduling problems, consisting in the assignment (over time) of scarce resources to customers. However, some special features characterize the charging scheduling problems treated in this chapter, namely the presence of time-varying energy costs and the integration of the charging apparatus within complex structures like a microgrid (that includes intermittent renewable energy sources). For this reason, it is necessary to develop specific and effective approaches for the considered scheduling problem.

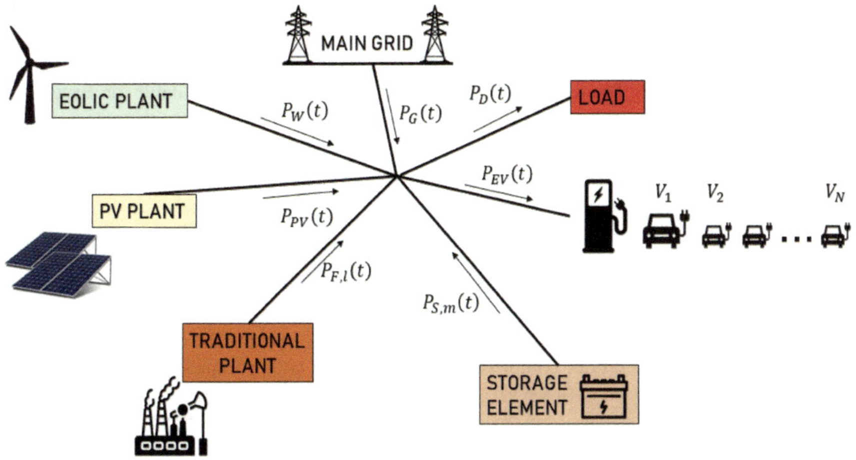

Fig. 4.2 The considered system

4.3 The Considered Model[2]

Figure 4.2 represents the system considered in this model formulation.

In order to define the mathematical model, the following sets are introduced:

- $\mathrm{EV} = \{1, \ldots, N\}$ set of electric vehicles;
- $\mathrm{FG} = \{1, \ldots, L\}$ set of fossil fuel generators.

The considered system consists of a microgrid in which the following elements are present:

- production from renewable energy sources: it is assumed that a forecast (in discrete time) of the power coming from the renewable source is available for any time slot in the optimization horizon;
- production from L energy sources fed by fossil fuel;
- a connection with the main grid, which allows bidirectional power flow between the main grid and the considered system;
- a charging station for EVs that may serve a single vehicle at a time; it is also assumed that the charging process is non-preemptive, that is, a given vehicle under charging *cannot* be disconnected (before the charging is completed) and reconnected later for the completion of its service; besides, every charging process must satisfy the energy request for any vehicle;
- a non-deferrable electric load; even in this case, a forecast (in discrete time) of this demand is assumed to be available for any time slot in the optimization horizon.

---

[2] The text in this section is © 2019 Elsevier. Based on [5], https://doi.org/10.1016/j.conengprac.2019.02.004; reproduced by permission.

Table 4.1 Known parameters for EVs

Symbol	Name	Description
rt_i	Release time	The time instant at which the vehicle becomes available for service (this may correspond to the forecasted arrival time at the service station)
dd_i	Due date	The time instant at which the service for the vehicle should be completed (according to the customer's preference)
dl_i	Deadline	The time instant at which the service must be completed (mandatory)
ER_i	Energy request	The amount of energy requested for the charging service
α_i	Penalty coefficient	The cost paid for a unit delay (with respect to the due date for the service completion, dd_i), per unit of energy requested. It is expressed in [€/(h kWh)]

For each vehicle V_i requiring the charging service, it is assumed that all information regarding arrival times, charging requests, and departure due times is known a priori. Such information is reported in Table 4.1 in detail.

The power flows in the system are

- $P_{\mathrm{EV}}(t)$, that is, the (monodirectional) power flow to the charging station; obviously, $P_{\mathrm{EV}}(t) = 0$ if at time instant t there is no vehicle under charge;
- $P_G(t)$, that is, the (bidirectional) power flow from the main grid to the microgrid; the active sign convention is adopted; namely, the power flow is assumed positive when power is drawn (bought) from the main grid (and negative when power is given (sold) to the main grid;
- $P_R(t)$, that is, the (monodirectional) power coming from renewables; it is given by $P_R(t) = P_W(t) + P_{\mathrm{PV}}(t)$ where the two terms on the right side represent the power generated by wind turbines ($P_W(t)$) and from photovoltaic power plants ($P_{\mathrm{PV}}(t)$);
- $P_{F,l}(t)$, that is, the (monodirectional) power flow from the lth energy source fed by fossil fuel to the microgrid;
- $P_D(t)$, that is, the (monodirectional) power flow representing the non-deferrable power demand;
- $P_S(t)$, that is, the (bidirectional) power flow from the energy storage element; here again, this flow is assumed positive when power is drawn from this element.

The SOC value at time instant t in the storage element will be denoted by $\mathrm{SOC}_S(t)$.

The service sequence is assumed to be given. In particular, the EVs are charged following the order of their arrival. Thus, the only decisions to be taken in the considered scheduling model are the starting time and the duration of the charging time intervals.

The energy request ER_i by vehicle V_i must be completely satisfied. Tardiness concerning the due date is introduced to penalize customer dissatisfaction. The presence of deadline constraints (regarding service completion times) is taken into account.

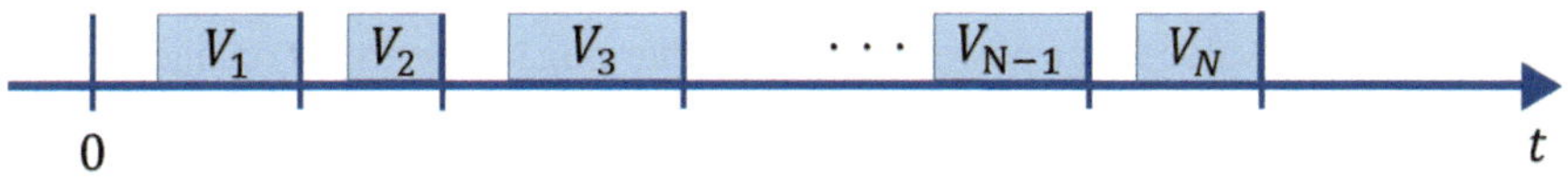

Fig. 4.3 The service sequence for the charging of vehicles

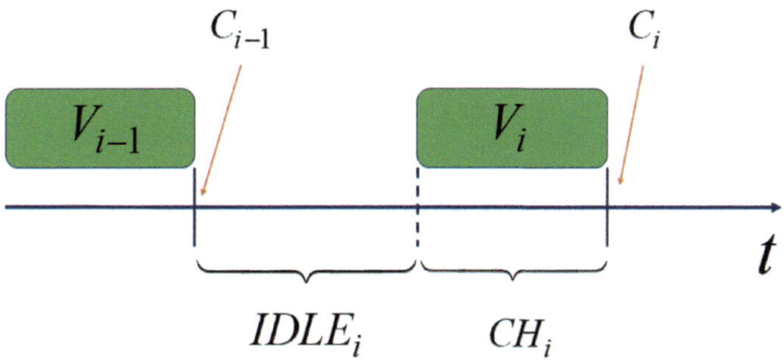

Fig. 4.4 The time intervals between two successive completion time instants. @ 2019 EUCA, Reproduced from [13], https://doi.org/10.23919/ECC.2019.8795999; with permission

The service sequence can thus be represented as in Fig. 4.3, where the lengths of the rectangles corresponding to the various services represent the charging service durations.

Let us denote as C_i the completion time instant of the charging of the ith vehicle. Then, the time interval between the charging completion time instants of two subsequent vehicles can be divided into (see Fig. 4.4).

- CH_i, that is, the charging time (interval) for the ith vehicle;
- IDLE_i, that is, the idle time interval *before* the charging of the ith vehicle.

Having chosen to adopt a discrete-event representation of the dynamics of the system, the only values of the state variables of the system (i.e., $\mathrm{SOC}_S(t)$) that must be evaluated are those corresponding to the completion of the services, that is, $\mathrm{SOC}_S(C_i)$ $i = 1, \ldots, N$, assuming that the initial state $\mathrm{SOC}_S(0)$ is given.

Then, the following *restrictive assumption* is introduced, with the objective of defining a parametric (instead of a functional) optimization problem:

the values of the power purchased/sold from/to the main grid $P_G(t)$, and the (costly) power generated by the lth fossil source $P_{F,l}(t)$, $l = 1, \ldots, L$ are kept at a constant value (resp. $P_{G,i,1}, P_{F,l,i,1}$) within each idle time interval $(C_{i-1}, C_{i-1} + \mathrm{IDLE}_i)$ and at another constant value (resp. $P_{G,i,2}, P_{F,l,i,2}$), within each charging time interval $(C_{i-1} + \mathrm{IDLE}_i, C_i)$, for $i = 1, \ldots, N$, $l = 1, \ldots, L$, C_0 being the initial time instant ($C_0 = 0$).

Although the value of the power flow to the charging station $P_{\mathrm{EV}}(t)$ cannot be kept constant within each charging time interval $(C_{i-1} + \mathrm{IDLE}_i, C_i)$, owing to the necessity of modeling the nonlinear behavior of the vehicles' battery, in a first formulation of the problem we will consider the average values $\overline{P}_{\mathrm{EV},i}$ $i = 1, \ldots, N$ as decision variables.

It is assumed that the forecasts of the renewable powers $P_W(t)$ and $P_{\mathrm{PV}}(t)$, as well as the power demand $P_D(t)$, and the selling/buying prices (SP(t) and BP(t)) to/from the main grid, are available and completely reliable for $t \geq 0$. That is to say, no uncertainty modeling is introduced. Thus, we introduce the function $P_{\mathrm{NL}}(t)$ which represents the net load and is defined as

$$P_{\mathrm{NL}}(t) = P_D(t) - P_W(t) - P_{\mathrm{PV}}(t) \quad i = 1, \ldots, N. \tag{4.1}$$

The power balance equation for any time instant t is given by

$$P_G(t) + \sum_{l=1}^{L} P_{F,l}(t) + P_S(t) = P_{\mathrm{NL}}(t) + P_{\mathrm{EV}}(t). \tag{4.2}$$

4.4 Piecewise Linear Battery Modeling

In real storage systems, the maximum injectable power depends on the state-of-charge, as described in the results of [14]. Specifically, considering a storage system, equipped with its battery management system, that receives commands from a central controller like an energy management system (EMS), it happens that, beyond a certain value of the state-of-charge, the maximum power that can be injected is lower than the rated value. This phenomenon can be (approximately) represented through a piecewise linear model as depicted in Fig. 4.5, where the feasible pairs (SOC, injectable power) are those inside or on the contour of the blue polygon. In this section, we consider only the charging of the batteries of the EVs.

The function represented in Fig. 4.5, providing the maximum injectable power for any value of the state-of-charge, can be expressed as

$$P_{\mathrm{EV},i}^{\mathrm{MAX}}(\mathrm{SOC}_{\mathrm{EV},i}) = \begin{cases} P_{\mathrm{EV},i}^{\mathrm{rated}} & \text{if } x_{\mathrm{EV},i} \leq a_{\mathrm{EV},i} \\ c_{\mathrm{EV},i}\mathrm{SOC}_{\mathrm{EV},i} + d_{\mathrm{EV},i} & \text{if } x_{\mathrm{EV},i} \geq a_{\mathrm{EV},i} \end{cases} \quad i = 1, \ldots, N \tag{4.3}$$

where of course parameters $a_{\mathrm{EV},i}, c_{\mathrm{EV},i}$, and $d_{\mathrm{EV},i}$ must satisfy the condition $c_{\mathrm{EV},i}a_{\mathrm{EV},i}+d_{\mathrm{EV},i} = P_{\mathrm{EV},i}^{\mathrm{rated}}$. Besides, $P_{\mathrm{EV},i}(1) = 0$ and thus it turns out $d_{\mathrm{EV},i} = -c_{\mathrm{EV},i}$.

Hence parameter $d_{\mathrm{EV},i}$ is simply given by

$$d_{\mathrm{EV},i} = \frac{P_{\mathrm{EV},i}^{\mathrm{rated}}}{1 - a_{\mathrm{EV},i}} \quad i = 1, \ldots, N. \tag{4.4}$$

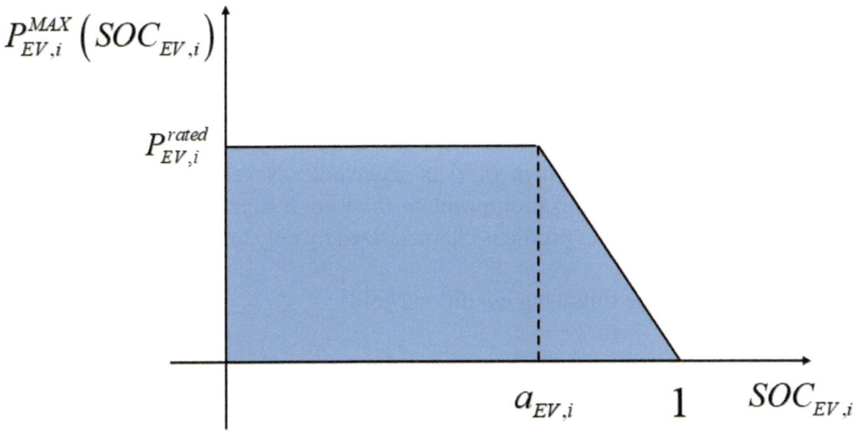

Fig. 4.5 Piecewise linear battery constraints (charging of EVs)

In this chapter, it is assumed that the charging process of a vehicle battery takes place according to the profile represented in Fig. 4.6 by the dark blue line, indicated in the following as the function:

$$P_{\mathrm{EV},i}(\mathrm{SOC}_{\mathrm{EV},i}) = \begin{cases} P_{\mathrm{EV},i}^{\mathrm{init}} & \text{if } \mathrm{SOC}_{\mathrm{EV},i} \leq b_{\mathrm{EV},i} \\ d_{\mathrm{EV},i}\left(1 - \mathrm{SOC}_{\mathrm{EV},i}\right) & \text{if } \mathrm{SOC}_{\mathrm{EV},i} \geq b_{\mathrm{EV},i} \end{cases} \quad i = 1, \ldots, N \quad (4.5)$$

where $P_{\mathrm{EV},i}^{\mathrm{init}}$ is the "nominal" power at which the charging process takes place, and parameter $b_{\mathrm{EV},i}$ satisfies the condition $d_{\mathrm{EV},i}\left(1 - b_{\mathrm{EV},i}\right) = P_{\mathrm{EV},i}^{\mathrm{init}}$, that is,

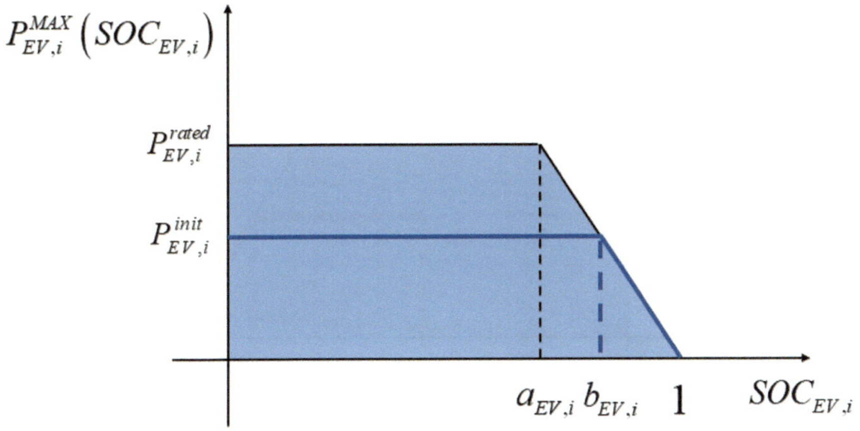

Fig. 4.6 The charging profile of the battery

$$b_{\mathrm{EV},i} = 1 - \frac{P_{\mathrm{EV},i}^{\mathrm{init}}}{d_{\mathrm{EV},i}} \quad i = 1, \ldots, N. \tag{4.6}$$

Of course, the choice of imposing that the charging process takes place according to a profile like the one represented in Fig. 4.6 is arbitrary. In fact, any profile lying below the piecewise function in (4.3) is acceptable. Nevertheless, as we aim at defining a parameter optimization problem (and not a functional one), we decide to impose that the charging profile is characterized by only one parameter, like $P_{\mathrm{EV},i}^{\mathrm{init}}$ in Fig. 4.6.

We assume that the following conditions hold:

$$\mathrm{SOC}_{\mathrm{EV},i}^{\mathrm{init}} \leq a_{\mathrm{EV},i} \quad i = 1, \ldots, N \tag{4.7}$$

$$\mathrm{SOC}_{\mathrm{EV},i}^{\mathrm{fin}} \geq b_{\mathrm{EV},i} \quad i = 1, \ldots, N \tag{4.8}$$

so that the actual charging profile is that represented in Fig. 4.7. The above assumption that (4.7) and (4.8) hold is equivalent to assuming that the requested charging service for any vehicle starts from relatively low values and finishes at relatively high values of the state-of-charge of its battery.

In the following, the value of $P_{\mathrm{EV},i}^{\mathrm{init}}$, that of course has to satisfy the condition $P_{\mathrm{EV},i}^{\mathrm{init}} \leq P_{\mathrm{EV},i}^{\mathrm{rated}}$, will be considered as a decision variable of the problem.

On this basis, if we assume that the (known) initial state-of-charge of the battery at the beginning of the charging process is $\mathrm{SOC}_{\mathrm{EV},i}^{\mathrm{init}}$ and that the (known) required state-of-charge, at the end of the charging process is $\mathrm{SOC}_{\mathrm{EV},i}^{\mathrm{fin}}$, the time needed for the charging process can be evaluated as

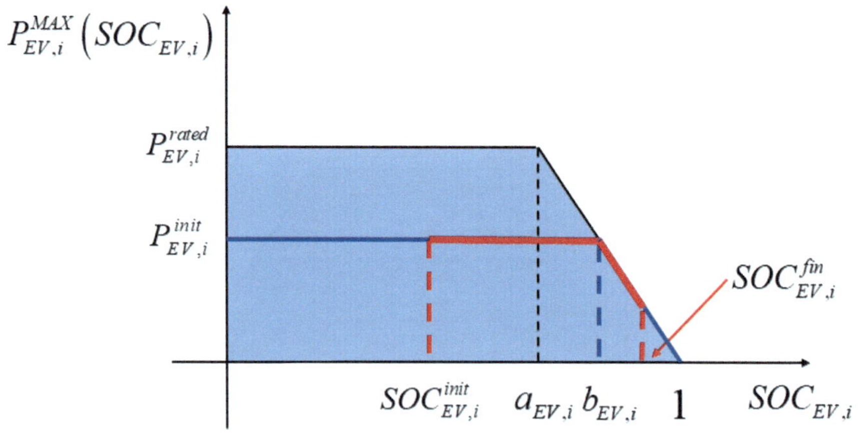

Fig. 4.7 The actual power profile of the battery (red line)

$$\mathrm{CH}_i = \mathrm{CAP}_{\mathrm{EV},i} \int_{\mathrm{SOC}_{\mathrm{EV},i}^{\mathrm{init}}}^{\mathrm{SOC}_{\mathrm{EV},i}^{\mathrm{fin}}} \frac{\mathrm{dSOC}}{P_{\mathrm{EV}}\left(\mathrm{SOC}_{\mathrm{EV},i}\right)} \quad i = 1, \ldots, N. \tag{4.9}$$

Given the above expression of $b_{\mathrm{EV},i}$, the condition (4.8) becomes

$$\mathrm{SOC}_{\mathrm{EV},i}^{\mathrm{fin}} \geq 1 - \frac{P_{\mathrm{EV},i}^{\mathrm{init}}}{d_{\mathrm{EV},i}} \quad i = 1, \ldots, N \tag{4.10}$$

that is,

$$P_{\mathrm{EV},i}^{\mathrm{init}} \geq d_{\mathrm{EV},i}\left(1 - \mathrm{SOC}_{\mathrm{EV},i}^{\mathrm{fin}}\right) \quad i = 1, \ldots, N. \tag{4.11}$$

The above inequality represents a constraint over the choice of the parameter $P_{\mathrm{EV},i}^{\mathrm{init}}$ that is induced by the value of the desired $\mathrm{SOC}_{\mathrm{EV},i}^{\mathrm{fin}}$.

Thus, the time interval required for charging can be evaluated as

$$\begin{aligned}
\mathrm{CH}_i\left(P_{\mathrm{EV},i}^{\mathrm{init}}\right) &= \mathrm{CAP}_{\mathrm{EV},i}\left(\int_{\mathrm{SOC}_{\mathrm{EV},i}^{init}}^{b_{\mathrm{EV},i}} \frac{\mathrm{dSOC}}{\eta_{\mathrm{EV}} P_{\mathrm{EV}}\left(\mathrm{SOC}_{\mathrm{EV},i}\right)} + \int_{b_{\mathrm{EV},i}}^{\mathrm{SOC}_{\mathrm{EV},i}^{\mathrm{fin}}} \frac{\mathrm{dSOC}}{\eta_{\mathrm{EV}} d_{\mathrm{EV},i}\left(1 - \mathrm{SOC}_{\mathrm{EV},i}\right)}\right) \\
&= \mathrm{CAP}_{\mathrm{EV},i}\left(\frac{b_{\mathrm{EV},i} - \mathrm{SOC}_{\mathrm{EV},i}^{\mathrm{init}}}{\eta_{\mathrm{EV}} P_{\mathrm{EV},i}^{\mathrm{init}}} - \frac{\ln\left(d_{\mathrm{EV},i} - d_{\mathrm{EV},i}\mathrm{SOC}\right)\Big|_{b_{\mathrm{EV},i}}^{\mathrm{SOC}_{\mathrm{EV},i}^{\mathrm{fin}}}}{\eta_{\mathrm{EV}} d_{\mathrm{EV},i}}\right) \\
&= \mathrm{CAP}_{\mathrm{EV},i}\left(\frac{1 - \frac{P_{\mathrm{EV},i}^{\mathrm{init}}}{d_{\mathrm{EV},i}} - \mathrm{SOC}_{\mathrm{EV},i}^{\mathrm{init}}}{\eta_{\mathrm{EV}} P_{\mathrm{EV},i}^{\mathrm{init}}} - \frac{\ln\left(d_{\mathrm{EV},i} - d_{\mathrm{EV},i}\mathrm{SOC}_{\mathrm{EV},i}^{\mathrm{fin}}\right) - \ln\left(P_{\mathrm{EV},i}^{\mathrm{init}}\right)}{\eta_{\mathrm{EV}} d_{\mathrm{EV},i}}\right) \\
&= \mathrm{CAP}_{\mathrm{EV},i}\left(\frac{1 - \frac{P_{\mathrm{EV},i}^{\mathrm{init}}}{d_{\mathrm{EV},i}} - \mathrm{SOC}_{\mathrm{EV},i}^{\mathrm{init}}}{\eta_{\mathrm{EV}} P_{\mathrm{EV},i}^{\mathrm{init}}} - \frac{\ln \frac{d_{\mathrm{EV},i}\left(1 - \mathrm{SOC}_{\mathrm{EV},i}^{\mathrm{fin}}\right)}{P_{\mathrm{EV},i}^{\mathrm{init}}}}{\eta_{\mathrm{EV}} d_{\mathrm{EV},i}}\right) \quad i = 1, \ldots N.
\end{aligned} \tag{4.12}$$

Note that in (4.12) the dependence of the duration of the charging service time on the initial input power $P_{\mathrm{EV},i}^{\mathrm{init}}$ has been put into evidence.

We recall that the following constraints must be fulfilled:

$$0 \leq P_{\mathrm{EV},i}^{\mathrm{init}} \leq P_{\mathrm{EV},i}^{\mathrm{rated}} \quad i = 1, \ldots, N. \tag{4.13}$$

4.5 The Proposed Optimization Approach

To formalize the considered discrete-event scheduling problem, we follow an approach structured as follows.

4.5.1 *Expressing the Length of the Charging Interval as a Function of the Initial Power*

Assuming that the charging profile in Fig. 4.7 has been selected, $P_{\mathrm{EV}}(t) = P_{\mathrm{EV},i}(t)$ can be represented as a function of the state-of-charge $\mathrm{SOC}_{\mathrm{EV},i}(t)$ as given by (4.5). In the following, the simplified notation

$$
\begin{aligned}
&P_{\mathrm{EV},i}(t) \rightarrow p(t)\\
&\mathrm{SOC}_{\mathrm{EV},i}(t) \rightarrow x(t)\\
&\mathrm{CAP}_{\mathrm{EV},i} \rightarrow \mathrm{CAP}\\
&d_{\mathrm{EV},i} \rightarrow d\\
&b_{\mathrm{EV},i} \rightarrow b\\
&P^{\mathrm{init}}_{\mathrm{EV},i} \rightarrow p^{\mathrm{init}}\\
&\mathrm{SOC}^{\mathrm{init}}_{\mathrm{EV},i} \rightarrow x^{\mathrm{init}}\\
&\mathrm{SOC}^{\mathrm{fin}}_{\mathrm{EV},i} \rightarrow x^{\mathrm{fin}}
\end{aligned}
$$

will be used.

The problem is that of integrating the differential equation

$$\dot{x} = \frac{\eta p(t)}{\mathrm{CAP}} \tag{4.14}$$

over the considered time interval. Note that η is the charging efficiency, and it is assumed to be constant during the charging process.

Integrating into the interval $x^{\mathrm{init}} \leq x \leq b$

In this interval

$$p(t) = p^{\mathrm{init}} \tag{4.15}$$

then

$$\dot{x} = \frac{\eta p^{\mathrm{init}}}{\mathrm{CAP}} \tag{4.16}$$

and thus, the solution is simply

$$x(t) = x^{\text{init}} + \frac{\eta p^{\text{init}}}{\text{CAP}} t. \tag{4.17}$$

We must determine the time instant t' at which $x(t)$ reaches the value b. That is,

$$b = x^{\text{init}} + \frac{\eta p^{\text{init}}}{\text{CAP}} t' \Rightarrow t' = \frac{b - x^{\text{init}}}{\eta p^{\text{init}}} \text{CAP}. \tag{4.18}$$

Since from (4.6) $b = 1 - \frac{p^{\text{init}}}{d}$, we obtain

$$t' = \frac{1 - \frac{p^{\text{init}}}{d} - x^{\text{init}}}{\eta p^{\text{init}}} \text{CAP}. \tag{4.19}$$

Of course, it must be $C_{i-1} + \text{IDLE}_i + t' \leq C_i$.
Thus, we can say that

$$p(t) = p^{\text{init}} \quad \text{for } C_{i-1} + \text{IDLE}_i \leq t \leq C_{i-1} + \text{IDLE}_i + t'. \tag{4.20}$$

Integrating into the interval $b \leq x \leq x^{\text{fin}}$

In this interval

$$p(t) = d(1 - x). \tag{4.21}$$

Thus, the differential equation that has to be integrated is

$$\dot{x} = \frac{\eta d(1 - x)}{\text{CAP}} \tag{4.22}$$

that is,

$$\dot{x} + \frac{\eta d}{\text{CAP}} x = \frac{\eta d}{\text{CAP}} \tag{4.23}$$

with the initial condition $x(t') = b$.
Solving the above differential equation (substituting $\frac{\eta d}{\text{CAP}}$ with α), we have

$$x(t) = 1\big(t - t'\big) + (b - 1)e^{-\frac{\eta d}{\text{CAP}}(t - t')} 1\big(t - t'\big). \tag{4.24}$$

Question When does the charging process end?
The conclusion of the charging process takes place when

$$x(t) = x^{\text{fin}} \tag{4.25}$$

then the charging process ends at time instant t'', where t'' satisfies

$$\begin{aligned} x^{\text{fin}} &= 1 + (b-1)e^{-\frac{\eta d}{\text{CAP}}(t''-t')} \\ \frac{x^{\text{fin}}-1}{b-1} &= e^{-\frac{\eta d}{\text{CAP}}(t''-t')} \\ (t''-t') &= -\frac{\text{CAP}}{\eta d}\ln\left(\frac{1-x^{\text{fin}}}{1-b}\right). \end{aligned} \tag{4.26}$$

But, of course, it must be $t'' = \text{CH}_i$. And thus,

$$\text{CH}_i = t' - \frac{\text{CAP}}{\eta d}\ln\left(\frac{1-x^{\text{fin}}}{1-b}\right). \tag{4.27}$$

Therefore, the expression of CH_i as a function of p^{init} is

$$\text{CH}_i = \frac{\text{CAP}}{\eta}\left[\frac{1-\frac{p^{\text{init}}}{d}-x^{\text{init}}}{p^{\text{init}}} - \frac{1}{d}\ln\left(\frac{d\left(1-x^{\text{fin}}\right)}{p^{\text{init}}}\right)\right] \tag{4.28}$$

Of course, the power profile is given by

$$p(t) = \begin{cases} p^{\text{init}} & C_{i-1} + \text{IDLE}_i \le t \le C_{i-1} + \text{IDLE}_i + t' \\ p^{\text{init}}e^{-\frac{\eta d}{\text{CAP}}(t-t')} & C_{i-1} + \text{IDLE}_i + t' \le t \le C_i. \end{cases} \tag{4.29}$$

4.5.2 *Determining the Minimum Length of the Charging Interval*

Owing to its physical meaning, CH_i is a monotonic nonincreasing function of the charging power (see Fig. 4.8), and its minimum coincides with the maximum value of p^{init}, i.e., p^{rated}. Note that the function in (4.28) is in fact a function of p^{init}.

Then, by substituting p^{rated} in p^{init} in (4.28), we obtain the minimum admissible duration for the charging interval CH_i, namely CH_i^{MIN}, given by

$$\text{CH}_i^{\text{MIN}} = \frac{\text{CAP}}{\eta}\left[\frac{1-\frac{p^{\text{rated}}}{d}-x^{\text{init}}}{p^{\text{rated}}} - \frac{1}{d}\ln\left(\frac{d\left(1-x^{\text{fin}}\right)}{p^{\text{rated}}}\right)\right]. \tag{4.30}$$

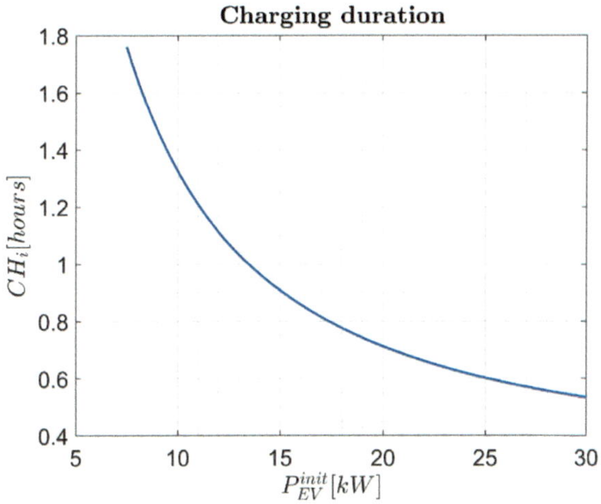

Fig. 4.8 Length of the charging interval according to the initial power

4.5.3 *Solving the Optimization Problem*

On the basis of the above discussion, an optimization problem can be considered, which provides the scheduling of the charging processes, determined considering average power flows. Specifically, let us consider the objective function which is composed of the following terms:

- the costs/benefits of energy taken/sold from/to the main grid during idle ($C_{\text{grid,IDLE},i}$) and charging ($C_{\text{grid,CH},i}$) time intervals;
- the cost of production by power plants from fossil fuel plants during idle ($C_{F,\text{IDLE},i}$) and charging ($C_{F,\text{CH},i}$) time intervals;
- the tardiness costs of the various vehicles ($C_{\text{tard},i}$).

Namely, the optimization objective is

$$\min \sum_{i=1}^{N} \{C_{\text{grid,IDLE},i} + C_{\text{grid,CH},i} + C_{F,\text{IDLE},i} + C_{F,\text{CH},i} + C_{\text{tard},i}\}. \tag{4.31}$$

To express the first two terms inside the summation, it is convenient to define

$$P_{G,i,1}^{+} = \max\{P_{G,i,1}, 0\} \quad i = 1, \dots, N \tag{4.32}$$

$$P_{G,i,1}^{-} = \max\{-P_{G,i,1}, 0\} \quad i = 1, \dots, N \tag{4.33}$$

and similarly

$$P_{G,i,2}^{+} = \max\{P_{G,i,2}, 0\} \quad i = 1, \dots, N \tag{4.34}$$

$$P_{G,i,2}^{-} = \max\{-P_{G,i,2}, 0\} \quad i = 1, \ldots, N \tag{4.35}$$

so that

$$P_{G,i,1} = P_{G,i,1}^{+} - P_{G,i,1}^{-} \quad i = 1, \ldots, N \tag{4.36}$$

$$P_{G,i,2} = P_{G,i,2}^{+} - P_{G,i,2}^{-} \quad i = 1, \ldots, N. \tag{4.37}$$

Thus, we have

$$C_{\text{grid,IDLE},i} = \left\{ \left[\int_{C_{i-1}}^{C_{i-1}+\text{IDLE}} \text{BP}(t)\text{d}t \right] P_{G,i,1}^{+} - \left[\int_{C_{i-1}}^{C_{i-1}+\text{IDLE}} \text{SP}(t)\text{d}t \right] P_{G,i,1}^{-} \right\} \quad i = 1, \ldots, N \tag{4.38}$$

$$C_{\text{grid,CH},i} = \left\{ \left[\int_{C_{i-1}+\text{IDLE}_i}^{C_i} \text{BP}(t)\text{d}t \right] P_{G,i,2}^{+} - \left[\int_{C_{i-1}+\text{IDLE}_i}^{C_i} \text{SP}(t)\text{d}t \right] P_{G,i,2}^{-} \right\} \quad i = 1, \ldots, N. \tag{4.39}$$

It is important to note that the terms $\int_{C_{i-1}}^{C_{i-1}+\text{IDLE}_i} \text{BP}(t)\text{d}t$, $\int_{C_{i-1}+\text{IDLE}_i}^{C_i} \text{BP}(t)\text{d}t$, $\int_{C_{i-1}}^{C_{i-1}+\text{IDLE}_i} \text{SP}(t)\text{d}t$, and $\int_{C_{i-1}+\text{IDLE}_i}^{C_i} \text{SP}(t)\text{d}t$ appearing in (4.38) and (4.39) are indeed known functions of C_i, C_{i-1}, and IDLE_i, since the availability and the reliability of a forecast of the patterns $\text{BP}(t)$ and $\text{SP}(t)$ have been assumed.

The production cost for each non-renewable plant l is given by the unit cost of production W_F [€/kWh] (assumed equal for each fossil power plant) multiplied by the power produced by that plant and the duration of the time interval (for idle and charging, respectively). The value of the overall production cost, during each time interval, is given by

$$C_{F,\text{IDLE},i} = \sum_{l=1}^{L} W_F P_{F,l,i,1} \text{IDLE}_i \quad i = 1, \ldots, N \tag{4.40}$$

$$C_{F,\text{CH},i} = \sum_{l=1}^{L} W_F P_{F,l,i,2} \text{CH}_i \quad i = 1, \ldots, N. \tag{4.41}$$

The tardiness cost is given by the tardiness in charging ($\text{tard}_i = \max\{C_i - \text{dd}_i, 0\}$) multiplied by the energy request for each vehicle ER_i [kWh] and a penalty coefficient α_i [€/kWh h] for unitary tardiness of the completion of charging service i. That is,

$$C_{\text{tard},i} = \alpha_i \text{tard}_i \text{ER}_i \quad i = 1, \ldots, N \tag{4.42}$$

As regards the constraints of the optimization problem, first we consider those representing the power balance, namely

$$\overline{P}_{G,i,1} + \overline{P}_{F,l,i,1} + \overline{P}_{S,i,1} = \overline{P}_{\mathrm{NL},i,1} \quad i = 1, \ldots, N \tag{4.43}$$

$$\overline{P}_{G,i,2} + \overline{P}_{F,l,i,2} + \overline{P}_{S,i,2} = \overline{P}_{\mathrm{NL},i,2} + \overline{P}_{\mathrm{EV},i} \quad i = 1, \ldots, N \tag{4.44}$$

where all the terms are defined as average values over the time intervals $(C_{i-1}, C_{i-1} + \mathrm{IDLE}_i)$ and $(C_{i-1} + \mathrm{IDLE}_i, C_i)$, respectively. Note that

$$\overline{P}_{\mathrm{NL},i,1} = \frac{1}{\mathrm{IDLE}_i} \int_{C_{i-1}}^{C_{i-1}+\mathrm{IDLE}_i} P_{\mathrm{NL}}(t)\mathrm{d}t \tag{4.45}$$

$$\overline{P}_{\mathrm{NL},i,2} = \frac{1}{\mathrm{CH}_i} \int_{C_{i-1}+\mathrm{IDLE}_i}^{C_i} P_{\mathrm{NL}}(t)\mathrm{d}t \quad i = 1, \ldots, N \tag{4.46}$$

are indeed known functions of C_i, C_{i-1}, and IDLE_i, since we have assumed the availability and the reliability of a forecast of the pattern $P_{\mathrm{NL}}(t)$.

Constraints defining the dynamics of the electrical storage must be taken into account:

$$x_{S,i,1}^{\mathrm{fin}} = x_{S,i,1}^{\mathrm{init}} + \eta_S^{\mathrm{char}} \overline{P}_{S,i,1}^{\mathrm{char}} \mathrm{IDLE}_i - \eta_S^{\mathrm{disch}} \overline{P}_{S,i,1}^{\mathrm{disch}} \mathrm{IDLE}_i \quad i = 1, \ldots, N \tag{4.47}$$

$$x_{S,i,2}^{\mathrm{fin}} = x_{S,i,2}^{\mathrm{init}} + \eta_S^{\mathrm{char}} \overline{P}_{S,i,2}^{\mathrm{char}} \mathrm{CH}_i - \eta_S^{\mathrm{disch}} \overline{P}_{S,i,2}^{\mathrm{disch}} \mathrm{CH}_i \quad i = 1, \ldots, N \tag{4.48}$$

$$x_{S,i,1}^{\mathrm{fin}} = x_{S,i,2}^{\mathrm{init}} \quad i = 1, \ldots, N \tag{4.49}$$

$$x_{S,i,2}^{\mathrm{fin}} = x_{S,i+1,1}^{\mathrm{init}} \quad i = 1, \ldots, N \tag{4.50}$$

where the power flows are distinguished among charging and discharging contributions. Owing to the choice of the optimization objective, for any time interval $\overline{P}_{S,i,1}^{\mathrm{char}} \left(\overline{P}_{S,i,2}^{\mathrm{char}}\right)$ and $\overline{P}_{S,i,1}^{\mathrm{disch}} \left(\overline{P}_{S,i,2}^{\mathrm{disch}}\right)$ cannot be jointly positive. The following constraints define the overall (average) power flow from/to the electrical storage

$$\overline{P}_{S,i,1} = \overline{P}_{S,i,1}^{\mathrm{char}} - \overline{P}_{S,i,1}^{\mathrm{disch}} \quad i = 1, \ldots, N \tag{4.51}$$

$$\overline{P}_{S,i,2} = \overline{P}_{S,i,2}^{\mathrm{char}} - \overline{P}_{S,i,2}^{\mathrm{disch}} \quad i = 1, \ldots, N. \tag{4.52}$$

Since the total energy request of each EV must be satisfied, we introduce the constraints

$$\overline{P}_{\mathrm{EV},i}\mathrm{CH}_i = \mathrm{ER}_i \quad i = 1, \ldots, N \tag{4.53}$$

where the energy request is given by

$$\mathrm{ER}_i = \left(x_{\mathrm{EV},i}^{\mathrm{fin}} - x_{\mathrm{EV},i}^{\mathrm{init}}\right)\mathrm{CAP}_{\mathrm{EV}} \quad i = 1, \ldots, N. \tag{4.54}$$

The deadlines must not be overcome

$$C_i \leq \mathrm{dl}_i \quad i = 1, \ldots, N. \tag{4.55}$$

Service cannot begin before the release time, and thus the difference between completion time and the time necessary for charging must be greater or equal to the release time:

$$C_i - \mathrm{CH}_i \geq \mathrm{rt}_i \quad i = 1, \ldots, N. \tag{4.56}$$

The idle time interval for each vehicle must be greater or equal to a minimum duration

$$\mathrm{IDLE}_i \geq \varepsilon \quad i = 1, \ldots, N. \tag{4.57}$$

Important constraints that must be introduced are those relevant to the minimum length of the charging interval (determined in Sect. 4.5.2), that is,

$$\mathrm{CH}_i \geq \mathrm{CH}_i^{\mathrm{MIN}} \quad i = 1, \ldots, N \tag{4.58}$$

where

$$\mathrm{CH}_i = C_i - C_{i-1} - \mathrm{IDLE}_i \quad i = 1, \ldots, N. \tag{4.59}$$

Besides, we must insert the constraints related to the definition of tardiness

$$\mathrm{tard}_i = \max\{C_i - \mathrm{dd}_i, 0\} \quad i = 1, \ldots, N. \tag{4.60}$$

We must also introduce bounds to limit some power flows (exchange with the external grid, power exchange with the storage systems, power production from traditional plants). Such constraints are, respectively,

$$-P_G^{\mathrm{MAX}} \leq \overline{P}_{G,i,1} \leq P_G^{\mathrm{MAX}} \quad i = 1, \ldots, N \tag{4.61}$$

$$-P_G^{\mathrm{MAX}} \leq \overline{P}_{G,i,2} \leq P_G^{\mathrm{MAX}} \quad i = 1, \ldots, N \tag{4.62}$$

$$-P_S^{\mathrm{MAX}} \le \overline{P}_{S,i,1} \le P_S^{\mathrm{MAX}} \quad i = 1, \ldots, N \tag{4.63}$$

$$-P_S^{\mathrm{MAX}} \le \overline{P}_{S,i,2} \le P_S^{\mathrm{MAX}} \quad i = 1, \ldots, N \tag{4.64}$$

$$P_F^{\mathrm{MIN}} \le \overline{P}_{F,l,i,1} \le P_F^{\mathrm{MAX}} \quad i = 1, \ldots, N, \quad l = 1, \ldots, L \tag{4.65}$$

$$P_F^{\mathrm{MIN}} \le \overline{P}_{F,l,i,2} \le P_F^{\mathrm{MAX}} \quad i = 1, \ldots, N, \quad l = 1, \ldots, L. \tag{4.66}$$

Additional constraints are relevant to the maximum and minimum state-of-charge

$$x^{\mathrm{MIN}} \le x_{S,i,1}^{\mathrm{init}} \le x^{\mathrm{MAX}} \quad i = 1, \ldots, N \tag{4.67}$$

$$x^{\mathrm{MIN}} \le x_{S,i,2}^{\mathrm{init}} \le x^{\mathrm{MAX}} \quad i = 1, \ldots, N \tag{4.68}$$

$$x^{\mathrm{MIN}} \le x_{S,i,1}^{\mathrm{fin}} \le x^{\mathrm{MAX}} \quad i = 1, \ldots, N \tag{4.69}$$

$$x^{\mathrm{MIN}} \le x_{S,i,2}^{\mathrm{fin}} \le x^{\mathrm{MAX}} \quad i = 1, \ldots, N. \tag{4.70}$$

Moreover, in order to ensure (4.11) we must impose

$$\overline{P}_{\mathrm{EV},i} \ge d\left(1 - x_i^{\mathrm{fin}}\right) \quad i = 1, \ldots, N. \tag{4.71}$$

Finally, all variables are understood to be non-negative, but $P_{G,i,1}$, $P_{G,i,2}$.

Remark 4.1 The formalization of the optimization problem has been carried out assuming the availability of the forecasts of net non-deferrable load and the buying–selling prices for the whole optimization horizon. In the problem statement, these forecasts are considered as completely reliable, and no uncertainty modeling is introduced. This fact gives rise to a predictive control scheme. Then, one can imagine repeatedly applying the proposed approach, each time conditioned by the most recent information about the system state and by the availability of the freshest predictions. For instance, one can state and solve a new problem at any service completion time (taking into account possible new service requests). In this way, a predictive (rolling horizon) discrete-event control scheme is applied.

4.5.4 Determining the Initial Charging Power

Once the problem in (4.31) s.t. (4.36)–(4.71) have been solved, it is possible to find the initial power by determining the relation between $\overline{P}_{\mathrm{EV},i}$ and the decision variable

p^{init}. This relation can be determined by integrating the function $p(t)$ in (4.29). Thus, we obtain

$$\begin{aligned}\overline{P}_{\text{EV},i}\text{CH}_i &= \int\limits_{C_{i-1}+\text{IDLE}_i}^{C_{i-1}+\text{IDLE}_i+t'} p^{\text{init}}\text{d}t + \int\limits_{C_{i-1}+\text{IDLE}_i+t'}^{C_i} \left[p^{\text{init}}e^{-\frac{\eta d}{\text{CAP}}(t-t')}\right]\text{d}t \\ &= p^{\text{init}}\left[t'\right] + p^{\text{init}}\left(-\frac{\text{CAP}}{\eta d}\right)\left[e^{-\frac{\eta d}{\text{CAP}}(\text{CH}_i-t')} - 1\right] \\ &= p^{\text{init}}\left[t' - \frac{\text{CAP}}{\eta d}\left(e^{-\frac{\eta d}{\text{CAP}}(\text{CH}_i-t')} - 1\right)\right].\end{aligned} \tag{4.72}$$

Then, the average power is determined by

$$\overline{P}_{\text{EV},i} = \frac{p^{\text{init}}\left[t' - \frac{\text{CAP}}{\eta d}\left(e^{-\frac{\eta d}{\text{CAP}}(\text{CH}_i-t')} - 1\right)\right]}{\text{CH}_i} \tag{4.73}$$

and substituting t' with $\frac{b-x^{\text{init}}}{\eta p^{\text{init}}}\text{CAP}$, we obtain

$$\overline{P}_{\text{EV},i} = \frac{p^{\text{init}}\left[\frac{1-\frac{p^{\text{init}}}{d}-x^{\text{init}}}{\eta p^{\text{init}}}\text{CAP} - \frac{\text{CAP}}{\eta d}\left(e^{-\eta d\left(\frac{\text{CH}_i}{\text{CAP}} - \frac{1-\frac{p^{\text{init}}}{d}-x^{\text{init}}}{\eta p^{\text{init}}}\right)} - 1\right)\right]}{\text{CH}_i}. \tag{4.74}$$

This equation can be solved numerically, and in this way the optimal value of p^{init} (for each vehicle i) is determined.

4.5.5 Determining the Optimal Power Flow Profiles to the Vehicles

Once the value of p^{init} is calculated, it is possible to find the profile of the charging power as described by (4.29). Figure 4.9 shows an example of how the result can appear.

4.5.6 Determining the Power Flows from/to the Storage Element

The same approach presented for the charging process of the EVs can be extended to the electrical storage. In fact, the difference between the two elements is given by the

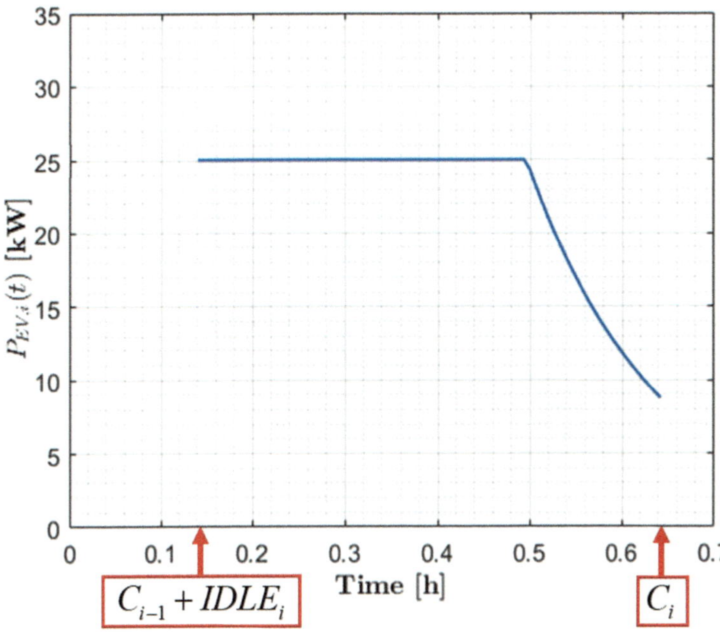

Fig. 4.9 Example of the time pattern of $p(t)$

energy request of the EVs which is mandatory while the storage is not constrained in that sense. Moreover, it is important to note that the storage is considered with the generator convention rather than the load one; thus the power is positive when the storage is discharged. The power profile must be within the region in Fig. 4.10.

Using an approach similar to the one presented before, and assuming that during a generic time interval the storage is charged or discharged, it is possible to obtain the power profile of the storage.

If the storage is discharged, the solution is very simple since we have

$$p_S(t) = \overline{P}_{S,i}^{\text{disch}}. \tag{4.75}$$

Instead, if the storage is charged, we have three possible outcomes for the state-of-charge of the storage at the edges of each time interval:

- both $x_{S,i,1}^{\text{init}}$ and $x_{S,i,2}^{\text{fin}}$ fall within the constant part (i.e., they are less or equal a_S);
- both $x_{S,i,1}^{\text{init}}$ and $x_{S,i,2}^{\text{fin}}$ fall within the descendent part (i.e., they are greater than a_S);
- $x_{S,i,1}^{\text{init}} \leq a_S \leq x_{S,i,1}^{\text{fin}}$ as in the case considered for the EVs.

In the first case, the power profile is equal to the average value

$$p_S(t) = \overline{P}_{S,i}^{\text{char}}. \tag{4.76}$$

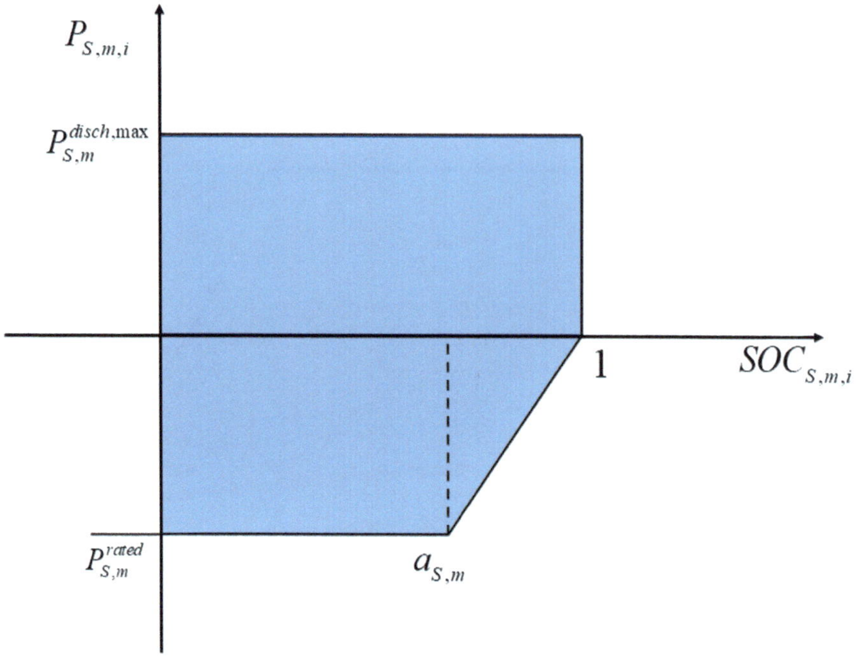

Fig. 4.10 Boundaries on the power exchange with the storage

In the second case the power is given by

$$p_S(t) = d_S(1 - x_{S,i,1}^{\text{init}})e^{\frac{\eta d_S}{\text{CAP}_S}t} \tag{4.77}$$

while, in the third case, we obtain

$$p_S(t) = \begin{cases} p_S^{\text{init,char}} & C_{i-1} + \text{IDLE}_i \leq t \leq C_{i-1} + \text{IDLE}_i + t'_S \\ p_S^{\text{init,char}} e^{\frac{\eta d_S}{\text{CAP}_S}(t - t'_S)} & C_{i-1} + \text{IDLE}_i + t'_S \leq t \leq C_i \end{cases} \tag{4.78}$$

where the initial power $p_S^{\text{init,char}}$ is determined by solving the following equation:

$$\overline{P}_{S,i}^{\text{char}} = \frac{p_S^{\text{init,char}} \left[\frac{1 - \frac{p_S^{\text{init,char}}}{d_S} - x^{\text{init}}}{\eta p_S^{\text{init,char}}} \text{CAP}_S + \frac{\text{CAP}_S}{\eta d} \left(e^{\eta d_S \left(\frac{\text{CH}_i}{\text{CAP}_S} - \frac{1 - \frac{p_S^{\text{init,char}}}{d} - x_{S,i-1}^{\text{init}}}{\eta p_S^{\text{init,char}}} \right)} - 1 \right) \right]}{\text{CH}_i}. \tag{4.79}$$

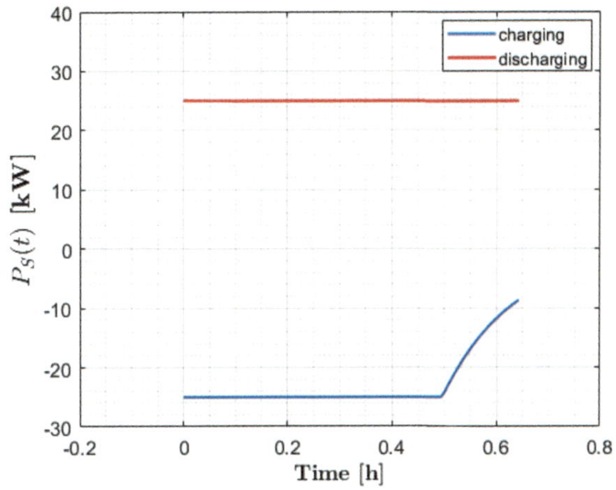

Fig. 4.11 Example of the charging (blue)/discharging (red) process of the storage

Note that here the initial (and final) state-of-charge of the storage is a solution of the optimization problem in Sect. 4.5.3, and also that in this case p_S^{init} and d assume negative values. An example of what can be obtained is presented in Fig. 4.11.

4.5.7 Establish the Power Exchanged with the Grid

We have simply established that

$$P_{F,l,i}(t) = \overline{P}_{F,l,i}. \tag{4.80}$$

Once the patterns $P_{F,i}(t)$, $P_{S,i}(t)$, and $P_{\text{EV},i}(t)$ have been determined, the pattern $P_{G,i}(t)$ can be determined as

$$P_{G,i}(t) = -P_{F,i}(t) - P_{S,i}(t) + P_{\text{NL},i}(t) + P_{\text{EV},i}(t). \tag{4.81}$$

Of course, it will surely result in

$$\int_{C_{i-1}}^{C_{i-1}+\text{IDLE}_i} P_{G,i,1}(t)\text{d}t = \overline{P}_{G,i,1}\text{IDLE}_i \tag{4.82}$$

$$\int_{C_{i-1}+\text{IDLE}_i}^{C_i} P_{G,i,2}(t)\text{d}t = \overline{P}_{G,i,2}\text{CH}_i \tag{4.83}$$

where $\overline{P}_{G,i,1}$ and $\overline{P}_{G,i,2}$ are exactly the values determined by solving the scheduling problem.

Remark 4.2 The only approximation introduced in this three-step approach regards the evaluation of the costs of type

$$\int_{C_{i-1}+\mathrm{IDLE}_i}^{C_i} \mathrm{BP}(t)P^{+}_{G,i,2}(t)\mathrm{d}t \tag{4.84}$$

through the approximation

$$\left[\int_{C_{i-1}+\mathrm{IDLE}_i}^{C_i} \mathrm{BP}(t)\mathrm{d}t\right]\overline{P}^{+}_{G,i,2}. \tag{4.85}$$

However, this approximation seems to be quite satisfying when $\mathrm{BP}(t)$ (and $\mathrm{SP}(t)$ in the case of sold power) is slowly time-varying (or constant) in the considered time interval $(C_{i-1} + \mathrm{IDLE}_i,\ C_i)$.

4.6 Application to a Case Study

In this section, the optimization model is solved for two different scenarios. The first one is similar to the one presented in Chap. 3 with the discrete-time model, and the number (N) of the considered EVs is 4. In the second scenario, the number of vehicles to be charged is larger ($N = 10$), and a higher value of the maximum charging power is considered.

The optimal schedule of production plants, storage systems, and EV charging is obtained by solving the optimization problem above introduced, by using the Lingo optimization tool (www.lindo.com, Global Solver option), on a PC Intel i7, 16 GB RAM.

4.6.1 Scenario I

Table 4.2 provides the values of the parameters relevant to the elements of the microgrid in the first considered scenario.

In Table 4.3 the data relevant to each EV are reported.

In the statement of the optimization problem, one of the most critical issues is how to express function $P_{\mathrm{NL}}(t)$. We assume that the forecast of the net load $P_{\mathrm{NL}}(t)$ is available over a whole day with a time discretization step equal to 15 min. Then,

Table 4.2 System parameters

Parameter	Value
x_{MAX}	80 [kWh]
x_{Min}	20 [kWh]
P_G^{MAX}	100 [kW]
P_S^{MAX}	30 [kW]
$P_{F,l}^{\mathrm{MAX}}$	30 [kW] 65 [kW]
$P_{F,l}^{\mathrm{MIN}}$	0 [kW]
$x_{m,0}$	60 [kWh]
CAP	100 [kWh]
η_{ch}	0.9 [–]
η_{disch}	1.1 [–]
B	1000 [–]
ε	0.08 [h]
C_F	0.2 [€/kWh]
C_0	0 [h]
$P_{\mathrm{EV}}^{\mathrm{rated}}$	10 [kW]
$a_{\mathrm{EV},i}$	0.8 [–]

Table 4.3 Scenario I: EVs' data

EV	1	2	3	4
$\mathrm{CAP}_{\mathrm{EV},i}$	22	25	31	22
$\mathrm{SOC}_{\mathrm{EV},i}^{\mathrm{init}}$	0.35	0.20	0	0.42
$\mathrm{SOC}_{\mathrm{EV},i}^{\mathrm{fin}}$	0.95	0.95	0.95	0.95
α_i	0.1	0.1	0.1	0.1
rt_i	1.25	2.5	4.5	8.75
dd_i	5.5	9	12.5	15
dl_i	7.5	12	15.75	17.5

in order to compute the above function, we may choose to interpolate the available forecasts via a polynomial function of suitable order. We have chosen an eighth-order polynomial

$$P_{\mathrm{NL}}(t) = a_8t^8 + a_7t^7 + a_6t^6 + a_5t^5 + a_4t^4 + a_3t^3 + a_2t^2 + a_1t + a_0.$$

Using the MATLAB® tool, the parameters of the above function are determined as provided in Table 4.4.

Table 4.4 Net load polynomial approximation parameters

Parameter	Value
a_8	1.07E−05
a_7	− 1.06E−03
a_6	4.09E−02
a_5	− 7.68E−01
a_4	7.20E+00
a_3	− 3.05E+01
a_2	4.77E+01
a_1	7.41E+01
a_0	1.07E−05

In Fig. 4.12, the original pattern of the net load $P_{\mathrm{NL}}(t)$ is represented, as well as the interpolating curve.

As for the net load, also in the case of the buying price, a polynomial function has been determined. In this case, the polynomial interpolation (see Fig. 4.13) is an eighth-order one, and the coefficient calculated by means of the previously introduced MATLAB® tool is reported in Table 4.5.

The results of this scenario are presented in Figs. 4.14 and 4.15.

From the results, it is possible to denote that the EVs are charged at the maximum available power, and the largest portion of the power demand is satisfied by the fossil plants. Since the energy cost function is higher in the last part of the day, the system finishes the charging process around 17 o'clock.

The computational time in this case is about 1–2 s.

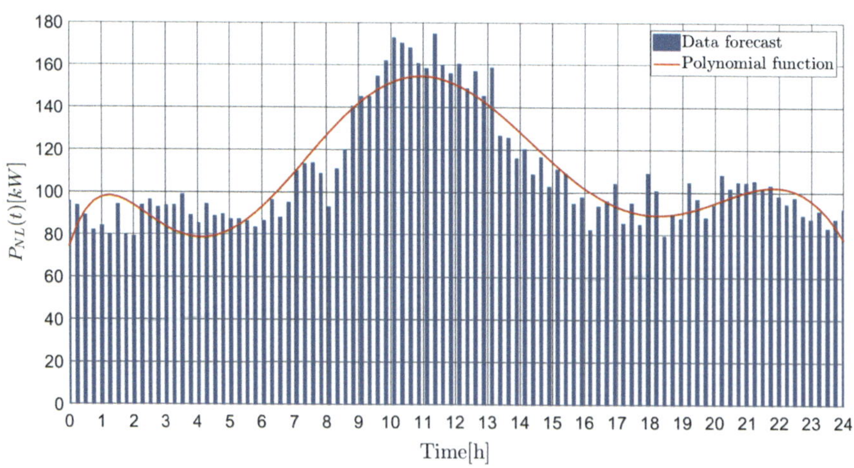

Fig. 4.12 Net load function and its polynomial approximation

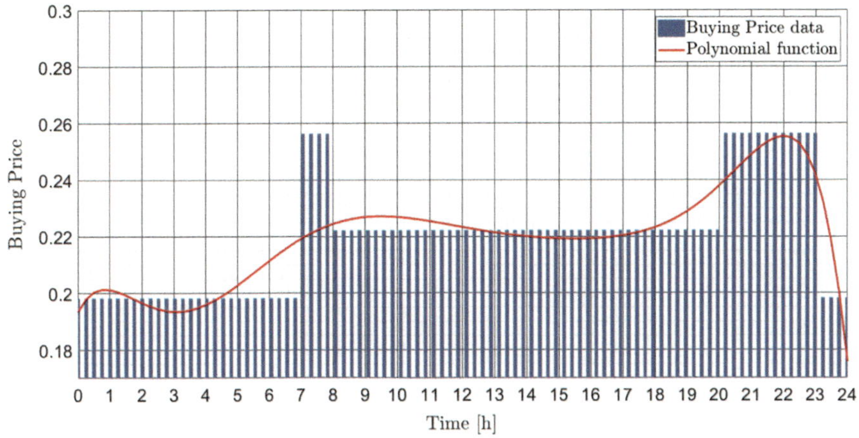

Fig. 4.13 Buying price function approximation

Table 4.5 Buying price polynomial approximation coefficients

Parameter	b_8	b_7	b_6	b_5	b_4	b_3	b_2	b_1	b_0
Value	– 1.35 E–09	1.26 E–07	– 4.85 E–06	9.90 E–05	– 1.13 E–03	7.06 E–03	– 2.08 E–02	2.25 E–02	1.94 E–01

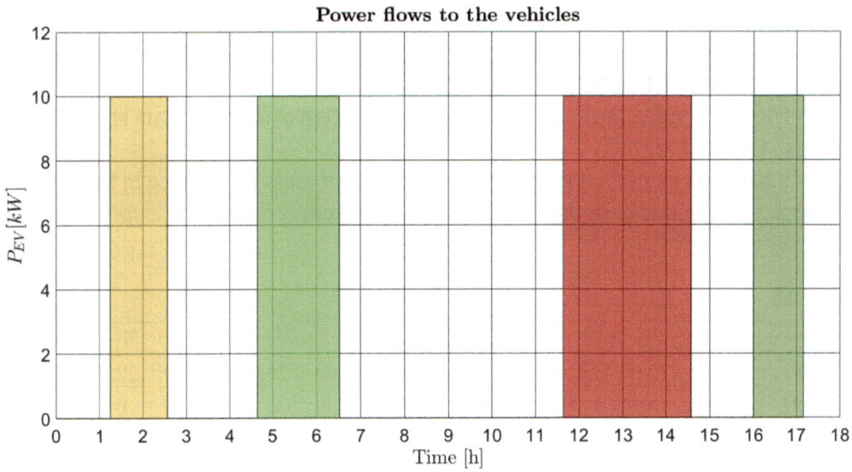

Fig. 4.14 Scenario I: power to the vehicles

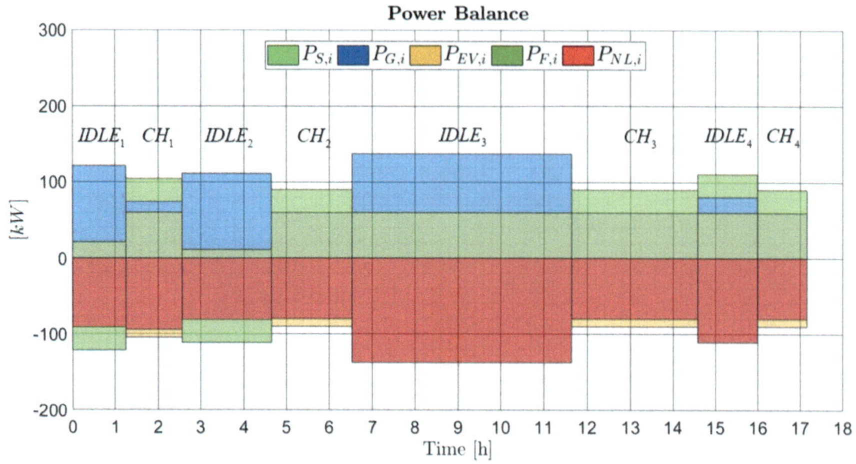

Fig. 4.15 Scenario I: power balance

4.6.2 *Scenario II*

The second scenario considered in this case study has the same values of the parameters in Table 4.2 but for the maximum power available for charging the EVs which is now doubled (i.e., $P_{\mathrm{EV}}^{\mathrm{rated}} = 20\,[\mathrm{kW}]$). The data relevant to the vehicles are presented in Table 4.6.

The polynomial approximations for the net load function and the buying price are the same considered in the previous scenario.

The results obtained from the optimization problem using these data are presented in Figs. 4.16 and 4.17.

As it happens in the first scenario, the charging process takes place in the first part of the day but in this case the IDLE intervals are reduced as much as possible. The

Table 4.6 Scenario II: EVs' data

EV	1	2	3	4	5	6	7	8	9	10
$\mathrm{CAP}_{\mathrm{EV},i}$	22.00	22.00	30.00	30.00	22.00	25.00	25.00	30.00	22.00	25.00
$\mathrm{SOC}_{\mathrm{EV},i}^{\mathrm{init}}$	0.35	0.20	0.00	0.42	0.55	0.60	0.30	0.15	0.42	0.50
$\mathrm{SOC}_{\mathrm{EV},i}^{\mathrm{fin}}$	0.95	0.95	0.95	0.95	0.95	0.95	0.95	0.95	0.95	0.95
α_i	0.10	0.10	0.10	0.10	0.10	0.10	0.10	0.10	0.10	0.10
rt_i	1.25	2.50	4.50	5.00	6.50	8.00	8.00	9.00	9.50	11.00
dd_i	5.50	9.00	12.00	12.50	13.00	16.00	16.00	19.00	20.00	22.00
dl_i	8.00	10.00	14.00	14.00	16.00	19.00	19.00	22.00	23.00	23.00

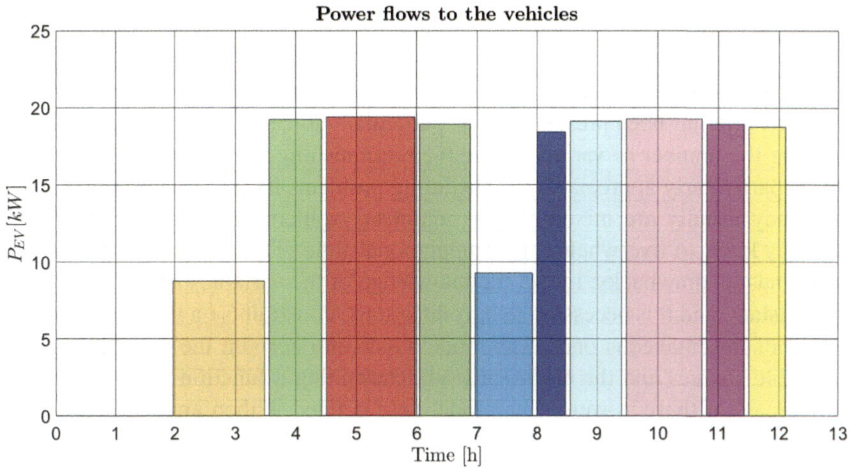

Fig. 4.16 Scenario II: power to the vehicles

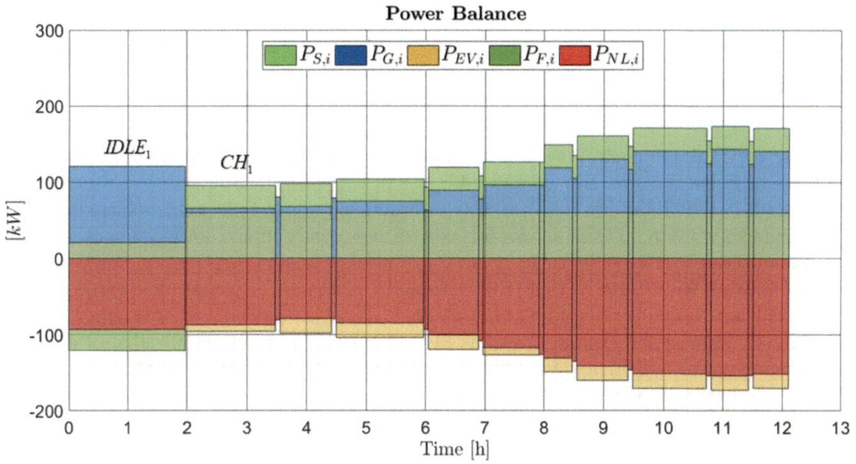

Fig. 4.17 Scenario II: power balance

power flow which charges the EVs is close to the maximum value in each interval except for vehicles 1 and 5. Fossil plants still provide a large portion of the power demand. The problem is solved in a few seconds (~ 9 s).

4.7 Conclusions

In this chapter, a discrete-event model for the charging of N electric vehicles (EVs) in a microgrid has been presented. This particular choice is due to the possibility of reducing the number of variables, and then diminishing the computational burden. This is particularly appreciable in scheduling problems like the one presented here, since they usually are mixed-integer nonlinear problems, and a large number of variables leads to overwhelming computational time. The adoption of this model clearly has its drawbacks; in fact the knowledge of the variables is only relevant to event instant, and it is necessary to introduce some constraints on the average values instead of instantaneous ones. The model takes into account the variability of the renewable sources and the electrical load determining a function which represents the forecast of these powers. The model here defined is then applied in two case studies with four EVs and ten EVs, respectively. The results of the problem show that the vehicles can be charged in the microgrid framework, and the electric load (both utilities and EVs) can be satisfied by means of some traditional plants and the main grid.

References

1. Miao L (2018) On controlling prioritized discrete event systems with real-time constraints. Discrete Event Dynamic Syst 28:427–447
2. Costa T, Pena P, Takahashi R (2018) SCO-Concat: a solution to a planning problem in flexible manufacturing systems using supervisory control theory and optimization techniques. J Control Autom Electr Syst 29(4):500–511
3. Jia QS, Wu J, Wu Z, Guan X (2018) Event-based HVAC control—a complexity-based approach. IEEE Trans Autom Sci Eng 15(4):1909–1919
4. Ferro G, Minciardi R, Parodi L, Robba M (2020) Discrete event optimization of a vehicle charging station with multiple sockets. Discrete Event Dyn Syst Theory Appl
5. Ferro G, Laureri F, Minciardi R, Robba M (2019) A predictive discrete event approach for the optimal charging of electric vehicles in microgrids. Control Eng Pract 86:11–23
6. Ferro G, Laureri F, Minciardi R, Robba M (2018) An optimization model for electrical vehicles scheduling in a smart grid. Sustain Energy Grids Netw 14:62–70
7. Pinedo M (2008) Scheduling. Springer, New York
8. French S (1987) Sequencing and scheduling. Wiley, Chichester
9. Brucker P (2007) Scheduling algorithms. Springer, Berlin, Heidelberg
10. Yang H, Pan H, Luo F, Qiu J, Deng Y, Lai M, Dong Z (2017) Operational planning of electric vehicles for balancing wind power and load fluctuations in a microgrid. IEEE Trans Sustain Energy 8(2):592–604
11. Kong F, Xiang Q, Kong L, Liu X (2016) On-line event-driven scheduling for electric vehicle charging via park-and-charge. In: 2016 IEEE real-time systems symposium (RTSS)
12. Di Giorgio, Liberati F, Canale S (2013) IEC 61851 compliant electric vehicle charging control in smartgrids. In: 21st Mediterranean conference on control and automation

13. Ferro G, Laureri F, Minciardi R, Robba M (2019) Optimal charging of electric vehicles in microgrids through discrete event optimization. In: 2019 18th European control conference (ECC), Naples, Italy, pp 916–921
14. Delfino F, Ferro G, Minciardi R, Robba M, Rossi M, Rossi M (2019) Identification and optimal control of an electrical storage system for microgrids with renewables. Sustain Energy Grids Netw 17:100183

Chapter 5
Optimal Charging in Smart Grids: A Discrete-Event Approach for Aperiodic Scheduling in Multi-socket Charging Stations

5.1 Introduction

As in Chap. 4, this chapter presents a discrete-event optimization problem for aperiodic scheduling but focuses attention on a more complex formalization that derives from the possibility of considering multiple sockets in a single charging station (CS). CSs may have different sockets, and thus different vehicles can be charged. However, this may create problems in meeting the deadlines for the recharge because, since the connection with the electrical grid has some limits in terms of maximum power: the charging time of the vehicles may be longer. Moreover, a general formalization of a multi-socket optimization problem with a high number of sockets also represents a charging park with many CSs all connected to the same electrical grid and with a single point of connection with the external grid [1, 2]. A similar approach has been presented in [3, 4], but the following features of the optimization problem have been added: the battery modeling is different and more detailed as regards efficiencies and nonlinear behavior; electric vehicles (EVs) are included in a microgrid with multiple components (fossil-fueled generation plants, loads, etc.); a post-processing analysis about the actual profile of the charging process in a continuous-time representation is presented. The chapter is organized as follows. Section 5.2 reports the system model, while details on the battery modeling (initial charging power, optimal power flows to EVs) are shown in Sects. 5.3 and 5.4. Optimal results for a specific case study are reported in Sect. 5.5, while conclusions are drawn in Sect. 5.6.

G. Ferro et al., *Optimization of Electric-Vehicle Charging*, Advances in Industrial Control, https://doi.org/10.1007/978-3-031-61917-5_5

5.2 The System Model

This chapter will extend the model presented in Chap. 4 to include the possibility of having a multi-socket charging station. In this formulation of the problem, we decided to avoid the introduction of idle intervals. Thus, we discretize the time axis as in Fig. 5.1, where C_i is the completion time of vehicle V_i.

To start with the formalization of our problem, we consider the following sets:

- $\text{EV} = \{1, \ldots, N\}$ the set of electric vehicles;
- $\text{FG} = \{1, \ldots, L\}$ the set of fossil fuel generators.

The considered system represents a microgrid in which the following elements can be identified:

- production from renewable energy sources: a forecast of the energy coming from each of these sources is assumed to be available over all time slots included within the optimization horizon;
- production from L energy sources fed by fossil fuel;
- a connection with the main grid, which allows bidirectional power flow;
- a charging station for EVs that serves at most N_{sock} vehicles at the same time; a vehicle under charging may be disconnected (before the charging is completed) from the charging station and reconnected later for the completion of service;
- a non-deferrable electric load; even for this load a forecasted demand pattern over the considered time horizon is assumed to be available.

It is assumed that, for each vehicle V_i to be recharged, information regarding arrivals, charging needs, and departure due times is known a priori. Such information is reported in Table 5.1.

For each time interval (C_{i-1}, C_i), $i = 1, \ldots, N$, we define the power flows:

- $\overline{P}_{\text{EV},i,j}$: the average power flow delivered to vehicle $V_j, j = i,\ i+1, \ldots, N$;
- $\overline{P}_{G,i}$: the average power flow from/to the grid;
- $\overline{P}_{\text{CS},i} = \sum\limits_{j=i}^{N} \overline{P}_{\text{EV},i,j}$: the average total power flow provided to the vehicles that are under charge in that time interval;
- $\overline{P}_{S,i}$: the average power flow from/to the storage;
- $\overline{P}_{\text{NL},i}$: the average net load;

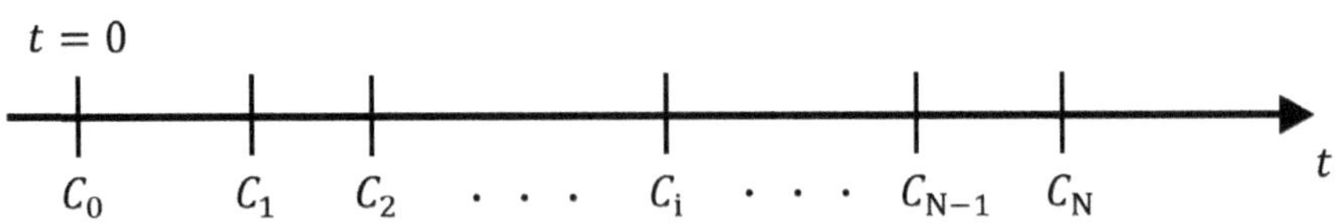

Fig. 5.1 Example of the discretization of the time axis. Reproduced from [3], released under a CC BY 4.0 license; https://doi.org/10.1007/s10626-020-00330-0

Table 5.1 Known parameters for EVs

Symbol	Name	Description
rt_i	Release time	The time instant at which the vehicle becomes available for service
dd_i	Due date	The time instant at which the service for the vehicle should be completed (according to the preference expressed by the customer)
dl_i	Deadline	The time instant at which the service must be completed
ER_i	Energy request	The amount of energy required for the charging service
α_i	Penalty coefficient	The cost paid for a unit delay (with respect to the due date dd_i), per unit of energy required. It is expressed in [€/(h kWh)]

- $P_{F,l,i}$: the power flow produced by the lth generation plant (constant over the considered time interval).

The considered system is similar to the one already presented except for the multiple sockets available in the charging station. The architecture depicted in Fig. 4.2 is changed to the one in Fig. 5.2.

We consider an approach like the previous chapter presented, composed of multiple steps. Since the detailed description of each step has already been presented, we do not repeat it here. Here we recall some of the nomenclature already defined (Table 5.2).

It is necessary to recall some of the previous chapter's ideas. The battery of the vehicles (as well as the storage) has a particular behavior since the maximum power injectable is not constant but a function of the state-of-charge. In this formalization, we approximate this limit with a function characterized by a constant (for low values of the state-of-charge) and a linear descending line. Following the approach presented

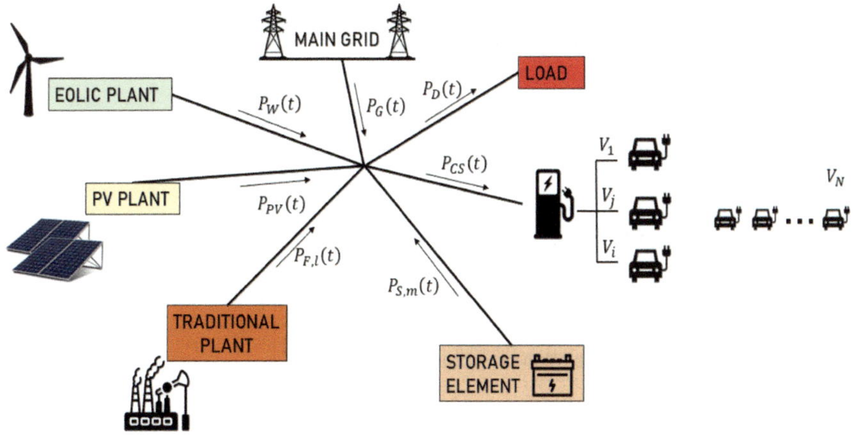

Fig. 5.2 The considered system; modified from Fig. 4.2

Table 5.2 Nomenclature from Chap. 4

Variable name	Description
p^{init}	Initial charging power of a generic EV
CAP	Battery capacity of a generic EV
$x^{\text{init}}(x^{\text{fin}})$	Initial (final) state-of-charge of a generic EV's battery
d,a	Parameters of the maximum power function
η	Efficiency of the charging process

in Sects. 4.4 and 4.5, we can write the equation which defines the relationship between the length of the charging interval CH_i^{CP} and the initial charging power p^{init}.

Note that CH_i^{CP} has the same meaning as the variable CH_i of Chap. 4, but in this case, since the charging process could last more than one interval, it is necessary to highlight some differences:

$$\text{CH}_i^{\text{CP}} = \frac{\text{CAP}}{\eta}\left[\frac{1-\frac{p^{\text{init}}}{d}-x^{\text{init}}}{p^{\text{init}}}-\frac{1}{d}\ln\left(\frac{d\left(1-x^{\text{fin}}\right)}{p^{\text{init}}}\right)\right]. \tag{5.1}$$

Once this function is determined, we can calculate the minimum length of the charging process by substituting the maximum value of p^{init}, i.e., p^{rated}, thus obtaining

$$\text{CH}_i^{\text{MIN}} = \frac{\text{CAP}}{\eta}\left[\frac{1-\frac{p^{\text{rated}}}{d}-x^{\text{init}}}{p^{\text{rated}}}-\frac{1}{d}\ln\left(\frac{d\left(1-x^{\text{fin}}\right)}{p^{\text{rated}}}\right)\right]. \tag{5.2}$$

Now it is possible to define the optimization problem.

5.3 The Optimization Problem[1]

An optimization problem is considered that provides the scheduling of the charging processes determined considering average power flows. Specifically, let us consider the objective function, which is composed of the following terms:

- the costs/benefits of energy taken/sold from/to the main grid during the charging ($C_{\text{grid},i}$) time interval;
- the cost of production by power plants from fossil fuel plants during the charging ($C_{F,i}$) time interval;
- the tardiness costs of the various vehicles ($C_{\text{tard},i}$).

Namely, the optimization objective is

[1] Developed from [3], https://doi.org/10.1007/s10626-020-00330-0; released under a CC BY 4.0 license.

$$\min \sum_{i=1}^{N} \left\{C_{\text{grid},i} + C_{F,i} + C_{\text{tard},i}\right\}. \tag{5.3}$$

Then, the different costs which compose the objective function are the following. The costs/benefits of energy taken/sold from/to the main grid are given by

$$C_{\text{grid},i} = \left\{\left[\int_{C_{i-1}}^{C_i} \text{BP}(t)\text{d}t\right]\overline{P}_{G,i}^{+} - \left[\int_{C_{i-1}}^{C_i} \text{SP}(t)\text{d}t\right]\overline{P}_{G,i}^{-}\right\}. \tag{5.4}$$

The cost relevant to the fossil fuel plants is given by

$$C_{F,i} = \sum_{l=1}^{L} W_F P_{F,l,i}\text{CH}_i \quad i = 1, \ldots, N \tag{5.5}$$

where W_F is the unit cost of the produced power (€/kWh).

The last cost to be considered is the one relevant to the tardiness in the charging process tard_i which is defined as

$$\text{tard}_i = \max\{C_i - \text{dd}, 0\} \quad i = 1, \ldots, N. \tag{5.6}$$

This value is multiplied by a penalty coefficient, namely α_i [€/kWh h] and the energy request of each vehicle ER_i [kWh], yielding

$$C_{\text{tard},i} = \alpha_i \text{tard}_i \text{ER}_i \quad i = 1, \ldots, N. \tag{5.7}$$

Having defined the objective function, let us introduce the problem's constraints. First, we introduce the power balance for each interval (C_{i-1}, C_i):

$$\overline{P}_{G,i} + \overline{P}_{F,l,i} + \overline{P}_{S,i} = \overline{P}_{\text{NL},i} + \overline{P}_{\text{CS},i} \quad i = 1, \ldots, N. \tag{5.8}$$

Note that the function $P_{\text{NL}}(t)$, already defined in (4.1) in the previous chapter, is given by the difference between the power load and the various renewable power forecasts such as PV and wind production.

The storage state equation is

$$x_{S,i} = x_{S,i-1} + \frac{\left(\eta_S^{\text{char}}\overline{P}_{S,i}^{\text{char}}\text{CH}_i - \eta_S^{\text{disch}}\overline{P}_{S,i}^{\text{disch}}\text{CH}_i\right)}{\text{CAP}_S} \quad i = 1, \ldots, N \tag{5.9}$$

where η_S^{char} and η_S^{disch} are charging (< 1) and discharging (> 1) efficiencies, respectively, CAP_S is the capacity of the battery, and the power flows are distinguished among charging and discharging contributions. Owing to the choice of the optimization objective, for any time interval $\overline{P}_{S,i}^{\text{char}}$ and $\overline{P}_{S,i}^{\text{disch}}$ cannot be jointly positive. Then, it is possible to write

$$\overline{P}_{S,i} = \overline{P}_{S,i}^{\text{disch}} - \overline{P}_{S,i}^{\text{char}} \quad i = 1, \ldots, N. \tag{5.10}$$

The energy request of each vehicle must be satisfied. This is ensured by constraints

$$\sum_{i=1}^{j} \overline{P}_{\text{EV},i,j}(C_i - C_{i-1}) = \text{ER}_j \quad j = 1, \ldots, N \tag{5.11}$$

where the following relationship provides the energy request

$$\text{ER}_i = \left(x_{\text{EV},i}^{\text{fin}} - x_{\text{EV},i}^{\text{init}}\right)\text{CAP}_{\text{EV}} \quad i = 1, \ldots, N. \tag{5.12}$$

It is now necessary to introduce a binary variable $y_{i,j}$ which is defined as

$$y_{i,j} = \begin{cases} 1 \text{ if vehicle } V_j \text{ is under charging in time interval } (C_{i-1}, C_i) \\ 0 \text{ otherwise} \end{cases}$$
$$i, j = 1, \ldots, N. \tag{5.13}$$

This variable must be used in the following constraints:

$$y_{i,j} \cdot K - \overline{P}_{\text{EV},i,j} \geq 0 \quad i, j = 1, \ldots, N. \tag{5.14}$$

The above constraints impose that when the power flow $\overline{P}_{\text{EV},i,j}$ is greater than zero, the binary variable $y_{i,j}$ is equal to 1. Note that K is the so-called "big M" constant whose value is arbitrarily chosen provided that it is considerably higher than the possible values of the variables and parameters appearing in the problem statement.

It is also necessary to impose that the charging process of a vehicle cannot start before the arrival time of the vehicle itself, that, in this formulation, corresponds to its release time. This is equivalent to imposing

$$y_{i+1,j} = 0 \quad \text{if } rl_j > C_i \quad i = 0, \ldots, N-1, j = 1, \ldots, N. \tag{5.15}$$

The above condition can also be stated as

$$\left(1 - y_{i+1,j}\right) K - \left(rl_j - C_i\right) \geq 0 \quad i = 0, \ldots, N-1, \quad j = 1, \ldots, N. \tag{5.16}$$

In this way, when $C_i < rl_j$, that is, when vehicle V_j is not yet ready for service at the end of time interval (C_{i-1}, C_i), $\left(1 - y_{i+1,j}\right)$ is forced to be equal to 1. This implies that $y_{i+1,j}$ is equal to 0, and for this reason vehicle V_j cannot be under charging in time interval (C_i, C_{i+1}).

As a matter of fact, one can note that writing this constraint as in (5.16) prevents the possibility of starting the service for a vehicle immediately after it has become ready for service. A vehicle ready for service before time instant C_i but later than time instant C_{i-1} cannot be put under charging, according to (5.16) before C_i, and thus has to wait for service even if some socket is free before C_i. Actually, this constraint reduces the responsiveness of the charging management policy. Nevertheless, this choice is forced by the necessity of reducing the number of time intervals (and thus of the decision variables).

The binary variables $y_{i,j}$ are also used to limit the number of vehicles that can be charged at the same time by using the following constraint:

$$\sum_{j \in N} y_{i,j} \leq N_{\text{sock}} \quad i = 1, \ldots, N. \tag{5.17}$$

In order to simplify the representation of the real charging profile (in the next section), it is important to add another assumption in the formalization. In fact, we suppose that the average power considered in the charging process of a vehicle V_j is the same for any time interval in which the charging process takes place. Thus, we can write

$$\overline{P}_{\text{EV},i,j} = y_{i,j} \overline{P}^{\text{CP}}_{\text{EV},j} \quad i, \; j = 1, \ldots, N \tag{5.18}$$

where $\overline{P}^{\text{CP}}_{\text{EV},j}$ is the constant average rate of charge.

Of course, the overall power delivered by the charging station is given by

$$\overline{P}_{\text{CS},i} = \sum_{j=i}^{N} \overline{P}_{\text{EV},i,j} \quad i = 1, \ldots, N. \tag{5.19}$$

Among the other constraints to be considered, there are those concerning the deadlines

$$C_i \leq \text{dl}_i \quad i = 1, \ldots, N \tag{5.20}$$

and those expressing the total time interval corresponding to the entire charging process of each vehicle, i.e., $\text{CH}_i^{\text{CP}} = \sum_{j=1}^{N} y_{i,j} \text{CH}_i$.

Besides, it must be, as in Chap. 4,

$$\text{CH}_i^{\text{CP}} \geq \text{CH}_i^{\text{MIN}} \quad i = 1, \ldots, N. \tag{5.21}$$

Then, we must take into account the bounds necessary to limit some power flows (exchange with the external grid, power exchange with the storage systems, power production from traditional plants, and power to the charging station). These bounds are, respectively,

$$-P_G^{\mathrm{MAX}} \le \overline{P}_{G,i} \le P_G^{\mathrm{MAX}} \quad i = 1, \ldots, N \tag{5.22}$$

$$-P_S^{\mathrm{MAX}} \le \overline{P}_{S,i} \le P_S^{\mathrm{MAX}} \quad i = 1, \ldots, N \tag{5.23}$$

$$P_F^{\mathrm{MIN}} \le \overline{P}_{F,l,i} \le P_F^{\mathrm{MAX}} \quad i = 1, \ldots, N, \;\; l = 1, \ldots, L \tag{5.24}$$

$$0 \le \overline{P}_{\mathrm{CS},i} \le P_{\mathrm{CS}}^{\mathrm{MAX}} \quad i = 1, \ldots, N. \tag{5.25}$$

Moreover, some constraints relevant to the maximum and minimum state-of-charge must be considered:

$$x^{\mathrm{MIN}} \le x_{S,i} \le x^{\mathrm{MAX}} \quad i = 1, \ldots, N. \tag{5.26}$$

Besides, in accordance with constraints (4.11) in Chap. 4, we must impose

$$\overline{P}_{\mathrm{EV},i}^{\mathrm{CP}} \ge d\left(1 - x_i^{\mathrm{fin}}\right) \quad i = 1, \ldots, N. \tag{5.27}$$

Finally, all variables are understood to be non-negative, but $\overline{P}_{G,i}$ and $\overline{P}_{S,i}$.

5.4 Determining the Initial Charging Power for Each Vehicle

Note that, in our formalization, even if the charging process can admit preemption, we suppose to have a charging profile characterized by the same average power over each time interval in which the charging process occurs. Once the value of the average power of the charging profile $\overline{P}_{\mathrm{EV},i}^{\mathrm{CP}}$ has been evaluated, we can determine the initial power for each vehicle.

To do this, as in the previous chapter, we must express the relation between $\overline{P}_{\mathrm{EV},i}^{\mathrm{CP}}$ and the decision variable p^{init}, integrating the function

$$p(t) = \begin{cases} p^{\mathrm{init}} & C_{\mathrm{init}} \le t \le C_{\mathrm{init}} + t' \\ p^{\mathrm{init}} e^{-\frac{\eta d}{\mathrm{CAP}}(t-t')} & C_{\mathrm{init}} + t' \le t \le C_i \end{cases} \quad i = 1, \ldots, N \tag{5.28}$$

where C_{init} is the instant in which the charging process starts. Then we obtain

$$\overline{P}_{\mathrm{EV},i}^{\mathrm{CP}} \mathrm{CH}_i^{\mathrm{CP}} = \int_{C_{\mathrm{init}}}^{C_{\mathrm{init}}+t'} p^{\mathrm{init}} \mathrm{d}t + \int_{C_{\mathrm{init}}+t'}^{C_i} \left[p^{\mathrm{init}} e^{-\frac{\eta d}{\mathrm{CAP}}(t-t')}\right] \mathrm{d}t$$

$$= p^{\text{init}}\left[t'\right] + p^{\text{init}}\left(-\frac{\text{CAP}}{\eta d}\right)\left[e^{-\frac{\eta d}{\text{CAP}}(\text{CH}_i^{\text{CP}}-t')} - 1\right]$$

$$= p^{\text{init}}\left[t' - \frac{\text{CAP}}{\eta d}\left(e^{-\frac{\eta d}{\text{CAP}}(\text{CH}_i^{\text{CP}}-t')} - 1\right)\right] \qquad i = 1, \ldots, N \tag{5.29}$$

and finally (by substituting t' with $\frac{b-x^{\text{init}}}{\eta p^{\text{init}}}\text{CAP}$), we obtain

$$\overline{P}_{\text{EV},i}^{\text{CP}} = \frac{p^{\text{init}}\left[\frac{1-\frac{p^{\text{init}}}{d}-x^{\text{init}}}{\eta p^{\text{init}}}\text{CAP} - \frac{\text{CAP}}{\eta d}\left(e^{-\eta d\left(\frac{\text{CH}_i^{\text{CP}}}{\text{CAP}} - \frac{1-\frac{p^{\text{init}}}{d}-x^{\text{init}}}{\eta p^{\text{init}}}\right)} - 1\right)\right]}{\text{CH}_i^{\text{CP}}} \qquad i = 1, \ldots, N. \tag{5.30}$$

Then the equation is solved, and the optimal value of p^{init} (for each vehicle i) is numerically determined.

5.5 Optimal Power Flow to the EVs

Once the calculation of each p^{init} has been carried out, by using the equation of $p(t)$ presented in (5.29), we can determine the power profile for each vehicle. Then, taking into account the values of both the charging intervals CH_i and the binary variables $y_{i,j}$, it is possible to split (if preemption occurs) the power profile along the time axis. In Fig. 5.3 an example is reported.

The figure shows how we consider the overall charging profile as "partitioned" in different time intervals. In this example, the third EV ($i = 3$) is charged in the intervals (C_0, C_1) and (C_2, C_3) which corresponds to CH_1 and CH_3 (red lines), the overall charging profile is represented by their sum, i.e., CH_3^{CP} (green lines).

As we have already pointed out, we are assuming in the problem statement that a charging profile (of a certain vehicle) is characterized by the same average power over each time interval in which the charging process occurs. Of course, this cannot be true if the overall charging interval is split into several time intervals, as, for instance, in Fig. 5.3. Actually, this may induce some underestimation of the real average power needed in some charging interval, and thus may affect the significance of constraints (5.19) providing the overall power flow from the charging station. However, we prefer to maintain the present formulation, since the alternative would be that of having a different average power for each charging time interval (for the same vehicle), and this would yield a significant complication of the overall optimization problem.

As presented in Chap. 4, the storage also has a behavior similar to the one described for the vehicles. The approach remains the same since the multi-socket extension is relevant only to the EVs.

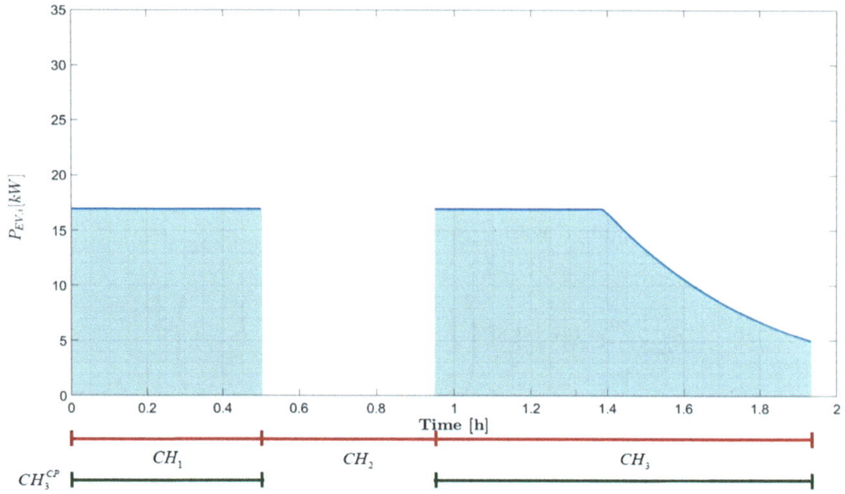

Fig. 5.3 Example of a charging profile divided into more time intervals

5.6 Application to a Case Study

The approach presented in the previous sections is tested on a case study where the charging station is supposed to have three sockets that can work simultaneously. Three different scenarios are presented to show how the solution is affected by an increase in the vehicles' energy request. The system parameters are reported in Table 5.3.

The data relevant to the vehicles are reported in Table 5.4.

The function $P_{\mathrm{NL}}(t)$ is represented via an eighth-order polynomial function as we have done in Chap. 4, and the coefficients are reported in Table 5.5.

The same approach is also adopted for the buying price with the coefficients presented in Table 5.6.

The results of the first scenario are presented in Figs. 5.4, 5.5 and 5.6.

In the Gantt diagram, it is possible to note that the first vehicle is the only one charged in the first interval since the others are not available due to their release time. Only the sixth vehicle is characterized by preemption, while the others are charged without interruptions. From the power balance, it is possible to note that a large portion of the load is represented by the net load component and that it is satisfied by the power purchased from the grid. The storage is almost entirely discharged in the first interval and then it remains constant. The overall cost result is 90.12 with a run time of about 15 s.

In the second scenario, the size of the EVs' batteries is increased by 5 kWh maintaining the same request of initial and final states-of-charge. The new values for $\mathrm{CAP}_{\mathrm{EV},i}$ are reported below (Table 5.7).

The resulting schedule of the charging processes is presented in the Gantt diagram in Fig. 5.7.

Table 5.3 System parameters

Parameter	Value
x^{MAX}	0.8 [–]
x^{MIN}	0.2 [–]
P_G^{MAX}	160 [kW]
P_{CS}^{MAX}	50 [kW]
P_S^{MAX}	30 [kW]
$P_{F,l}^{\mathrm{MAX}}$	65 [kW]
$P_{F,l}^{\mathrm{MIN}}$	0 [kW]
$x_{S,1,1}^{\mathrm{init}}$	0.6 [–]
CAP	100 [kWh]
η_{ch}	0.9 [–]
η_{disch}	1.1 [–]
K	200 [–]
ε	0.08 [h]
W_F	0.2 [€/kWh]
C_0	0 [h]
$P_{\mathrm{EV}}^{\mathrm{rated}}$	20 [kW]
$a_{\mathrm{EV},i}$	0.8 [–]
α_i	0.1 [–]

Table 5.4 Scenario I: EVs' data

EV	1	2	3	4	5	6	7	8
$\mathrm{CAP}_{\mathrm{EV},i}$	22	30	30	22	22	30	30	22
$\mathrm{SOC}_{\mathrm{EV},i}^{\mathrm{init}}$	0.35	0.20	0.30	0.40	0.30	0.45	0.6	0.4
$\mathrm{SOC}_{\mathrm{EV},i}^{\mathrm{fin}}$	0.95	0.95	0.95	0.95	0.95	0.95	0.95	0.95
rt_i	0	1	1	3	3	3	4	4
dd_i	2	2	3	3	6	6	6	6
dl_i	3	4	4	6	8	8	8	8

Table 5.5 Net load polynomial approximation parameters

Parameter	Value
a_7	1.07E−05
a_6	− 1.06E−03
a_5	4.09E−02
a_4	− 7.68E−01
a_3	7.20E+00
a_2	− 3.05E+01
a_1	4.77E+01
a_0	7.41E+01

Table 5.6 Buying price polynomial approximation coefficients

Parameter	b_8	b_7	b_6	b_5	b_4	b_3	b_2	b_1	b_0
Value	– 1.35 E–09	1.26 E–07	– 4.85 E–06	9.90 E–05	– 1.13 E–03	7.06 E–03	– 2.08 E–02	2.25 E–02	1.94 E–01

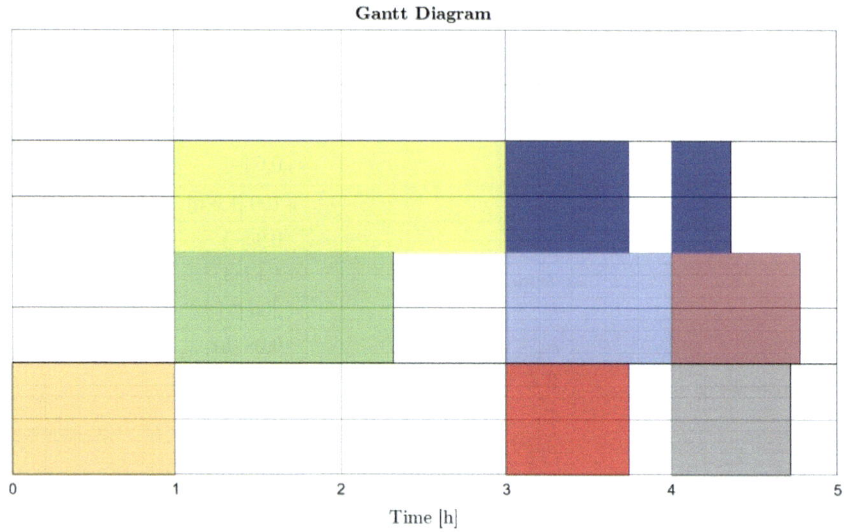

Fig. 5.4 Scenario 1: Gantt diagram

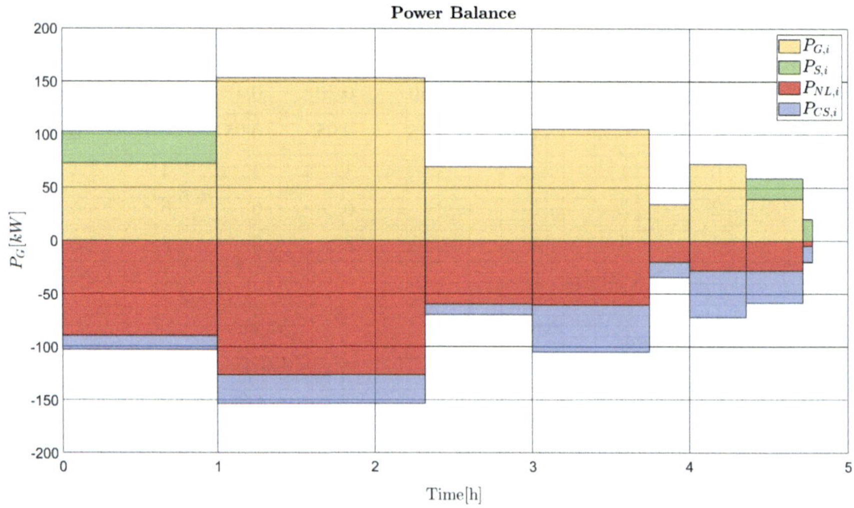

Fig. 5.5 Scenario 1: power balance

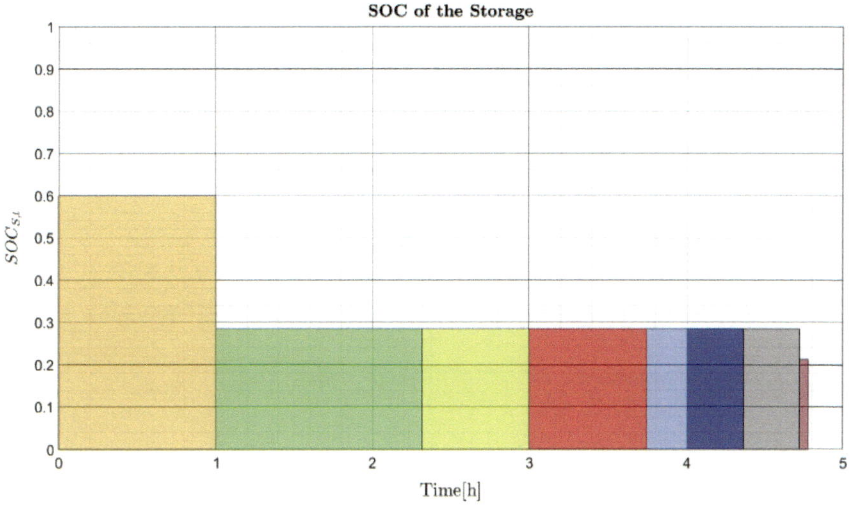

Fig. 5.6 Scenario 1: state-of-charge of the battery

Table 5.7 Scenario II: EVs' data

EV	1	2	3	4	5	6	7	8
$CAP_{EV,i}$	27	35	35	27	27	35	35	27

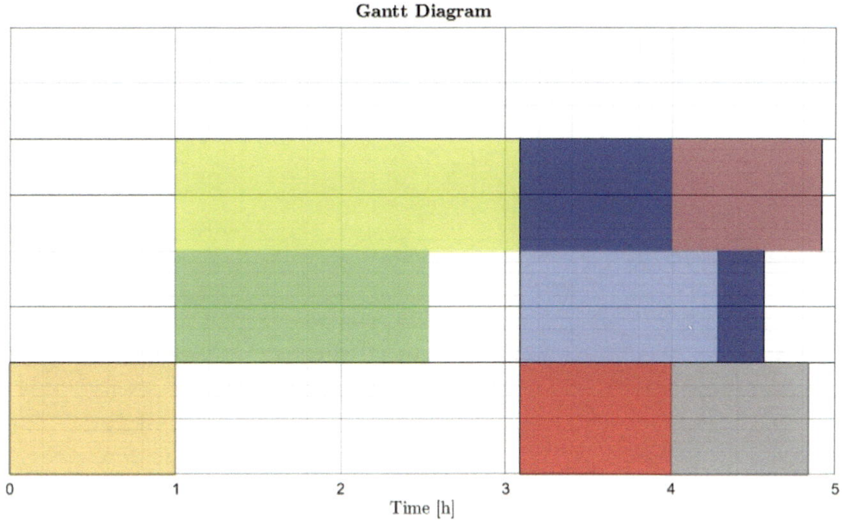

Fig. 5.7 Scenario 2: Gantt diagram

It is clear how the first EV remains the only one to be charged in the first interval due to the constraints on the release time, the solution for EVs 2 and 3 remains similar, and the remaining vehicles must be charged taking up all the sockets.

The storage is discharged in the first interval (see Fig. 5.8) and the main difference in the power balance (Fig. 5.9) is the production of some power by means of the fossil fuel plants. The total cost is 108.15 and the run time is 16 s.

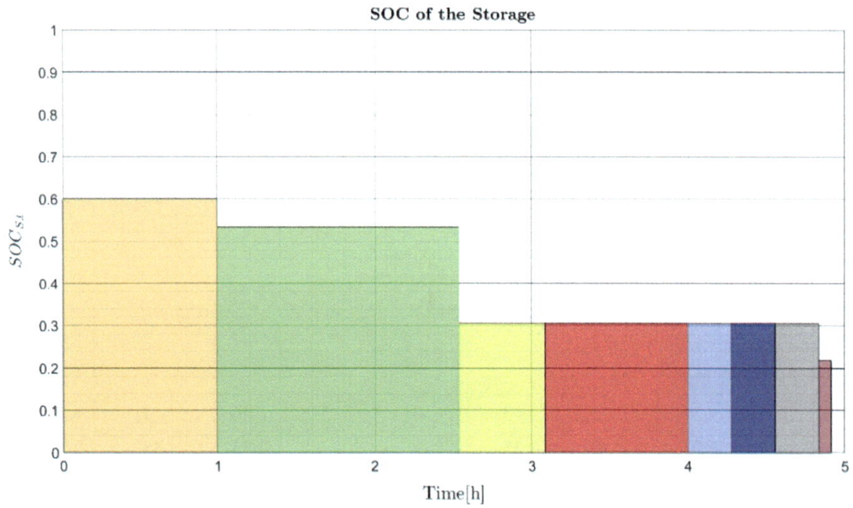

Fig. 5.8 Scenario 2: state-of-charge of the battery

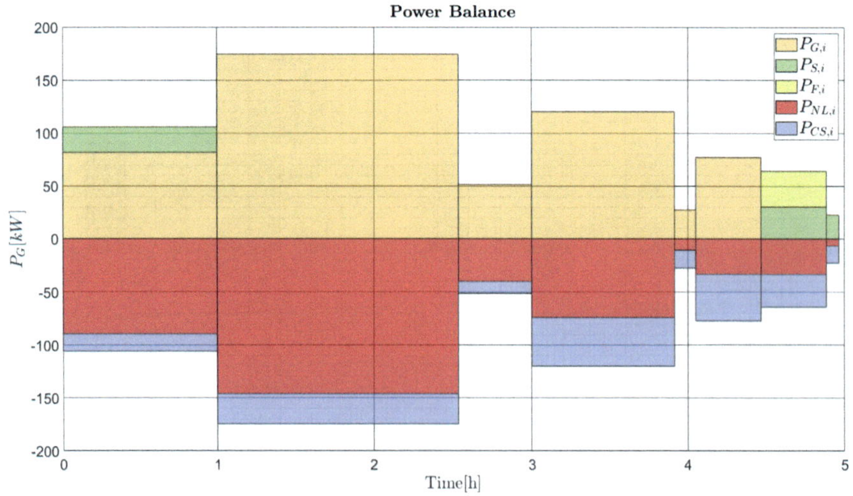

Fig. 5.9 Scenario 2: power balance

Table 5.8 Scenario III: EVs' data

EV	1	2	3	4	5	6	7	8
$CAP_{EV,i}$	32	40	40	32	32	40	40	32

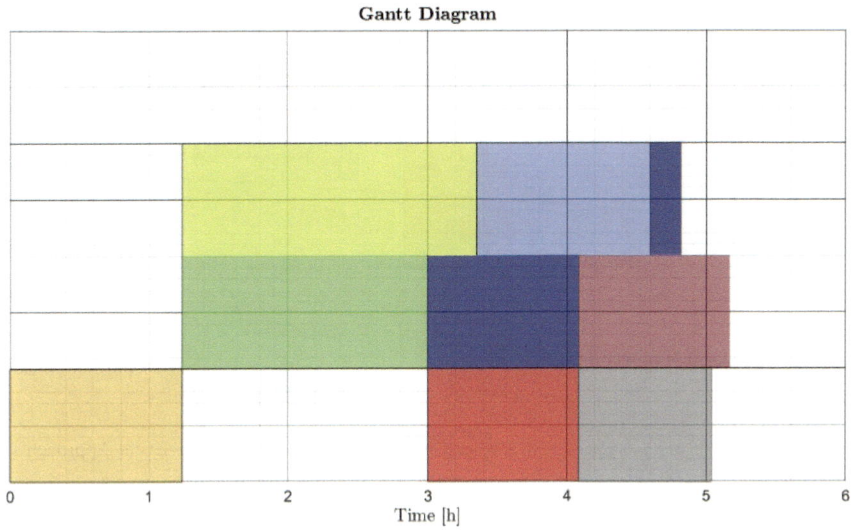

Fig. 5.10 Scenario 3: Gantt diagram

In the third scenario, a further increase in the batteries' size (i.e., in the energy request) has been considered. The new values are reported in Table 5.8.

In the following, the results are presented in Figs. 5.10, 5.11 and 5.12.

Figures 5.10, 5.11 and 5.12 show that the charging process ends a bit later than the previous scenarios. Moreover, there is a slightly higher production of fossil plants. The charge of some vehicles is also moved up because of the larger amount of power to be delivered before the deadline. The total cost is equal to 131.02 € and the run time is about 35 s.

5.7 Conclusions

In this chapter, an extension to the model described in Chap. 4 has been presented. In this case, we consider CSs that may have multiple sockets, and thus different vehicles can be charged in parallel. Moreover, a general formalization of a multi-socket optimization problem with a high number of sockets also represents a charging park with many CSs all connected to the same electrical grid and with a single point of connection with the external grid. The optimization model is applied to a case study where the charging station is supposed to have three sockets that can work

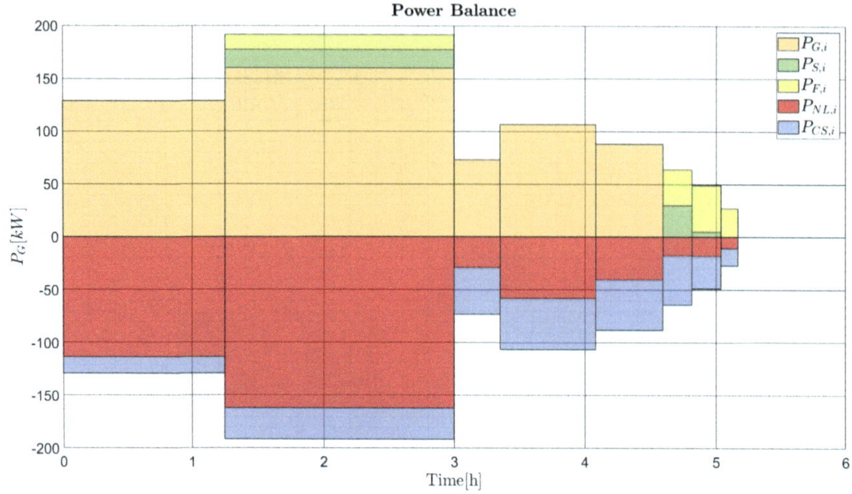

Fig. 5.11 Scenario 3: power balance

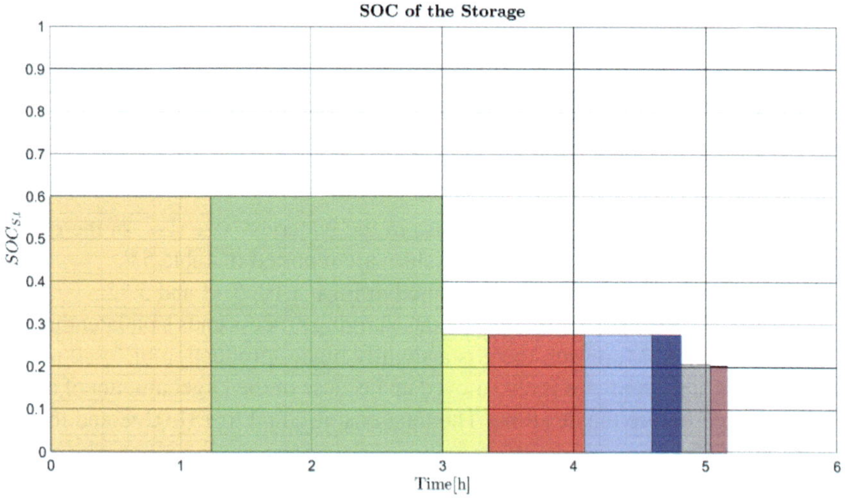

Fig. 5.12 Scenario 3: state-of-charge of the battery

simultaneously. Three different scenarios are presented to show how the solution is affected by an increase in the vehicles' energy requests.

References

1. Gough M, Lotfi M, Santos S, Espassandim H, Shafie-khah M, Catalao J (2021) Rooftop photovoltaic parking lots to support electric vehicles charging: a comprehensive survey. Int J Electr Power Energy Syst 133
2. Alinejad M, Rezaei O, Kazemi A, Bagheri S (2021) An optimal management for charging and discharging of electric vehicles in an intelligent parking lot considering vehicle owner's random behaviors. J Energy Storage 35
3. Ferro G, Minciardi R, Parodi L, Robba M (2021) Discrete event optimization of a vehicle charging station with multiple sockets. Discrete Event Dyn Syst: Theory Appl 31(2):219–249
4. Ferro G, Laureri F, Minciardi R, Robba M (2019) A predictive discrete event approach for the optimal charging of electric vehicles in microgrids. Control Eng Pract 86:11–23

Chapter 6
Optimal Charging in Smart Grids: A Discrete-Event Approach for Periodic Scheduling

6.1 Introduction

In this chapter, the optimal scheduling of the charging of EBs (Electric Buses) in a depot is considered. EBs are, in fact, EVs with larger sizes and are usually intended for public transportation purposes. The charging process of EBs is the same as the other EVs but for the size of the battery and, consequently, the time needed to be charged. An interesting survey on EBs is proposed by [1], where the authors analyze important components, existing research topics, and open issues of EBs. A substantial issue is related to the fact that they are usually located in very large depots, and, in some cases, they are requested to be charged in a specific time horizon. This can be very stressful for the grid and must be carefully managed. In recent years many studies about this topic have been proposed. Authors in [2] present a model for the exact route scheduling optimization considering the charging scheduling of a heterogeneous fleet. The authors try to reduce the number of buses charging simultaneously, imposing a grid limit, and find that heterogenous fleets can lead to advantages and the combined route and charging schedule. Another study that focuses on the EBs' scheduling is presented by [3]; the authors present a multi-objective bi-level model for charging and scheduling a fleet of EBs. Due to their size and well-defined operating schedule, they can represent a resource in some applications like the V2G as presented by [4]. A novel framework for joint optimal allocation and scheduling for EBs with V2G regulation service is studied.

Some works have highlighted the intrinsic periodic behavior of EBs schedules and the necessity to schedule the recharge taking into account this periodic behavior [5], which is indeed the main difference concerning other kinds of approaches for the scheduling of EVs [6, 7]. However, some uncertainties should be considered and they are particularly valid for the case of big EVs: efficiency of the batteries that diminishes with time, state-of-health of the battery, traffic conditions, number of passengers, and energy consumption that is strongly dependent on drivers' modes.

G. Ferro et al., *Optimization of Electric-Vehicle Charging*, Advances in Industrial Control, https://doi.org/10.1007/978-3-031-61917-5_6

Indeed, there is a huge effort to simulate energy consumption in literature. A simulation tool to evaluate electric buses (EBs) under different operating conditions has been developed in [8]. To determine the optimal battery capacity for electric buses, the authors in [9] use a MATLAB®-based electric vehicle simulator (ADVISOR). Furthermore, [10] presents a cost–benefit analysis of hybrid and electric city buses in fleet operation, taking into account energy consumption through extensive simulations over different bus routes. The study in [11] addresses the planning problem of fast charging stations within an electric bus system considering the energy consumption uncertainty of buses and using a robust optimization approach. A correct evaluation of consumption is particularly important in this case also for planning fast charging stations over the territory. In [12], the study aims to optimize the planning of both transportation systems and power grids. They put forward a spatial–temporal model that identifies the locations and sizes of e-bus charging stations, and outlines the strategies for multiple infrastructure stages. As indicated in reference [13], linking traffic/transportation and electrical networks is essential in determining the appropriate siting and sizing of charging stations. The study presented in [14] shows the architecture and implementation of a model created to optimize the allocation of electric bus charging infrastructure in urban areas. The model was put to the test using the Stockholm bus network.

In this chapter, a scheduling problem concerning the optimal charging of the EBs is considered within a discrete-event setting.

The considered system consists of a depot including the following elements (see Fig. 6.1): a connection to the main grid, an electrical storage element, a unique charging station, and N electric buses to be charged. The Vehicle-to-Grid mode is not allowed.

It is important to note that the electrical storage element has different efficiencies depending on the mode of operation (charging/discharging). Energy requests are calculated on the basis of the path and vehicle characteristics, by the use of a model

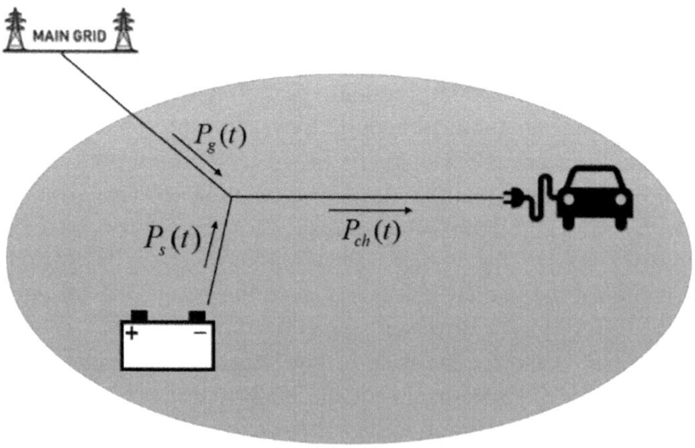

Fig. 6.1 The considered system

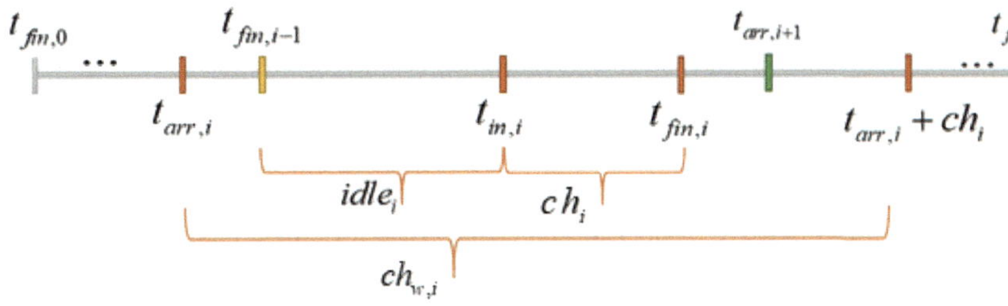

Fig. 6.2 The considered time windows for the *i*-th vehicle

for energy consumption. In the following, several sensitivity analyses have been performed to evaluate the influence of this parameter on the optimal solution.

6.2 The Optimization Problem[1]

The system model includes the following power flows, that are functions of time t:

- $P_g(t)$: the power flow from the main grid;
- $P_{ch}(t)$: the power flow to the bus in charge;
- $P_s(t)$: the power flow from/to the storage unit (active sign convention is used, as in other cases).

Each bus has an integer index $i = 1, \ldots, N$ assigned according to the arrival order. The following information is known a priori for each bus:

- *Arrival time* ($t_{\mathrm{arr},i}$): the instant when the charging of the i-th bus can start;
- *Energy consumption* (EC_i): the energy requested to charge the i-th bus (the request must be completely satisfied);
- *Available charging window* ($\mathrm{ch}_{w,i}$): time interval in which charging service for the i-th bus can take place.

The service times are controllable and are included within the set of the decision variables of the problem. Figure 6.2 represents the time windows concerning the i-th bus.

The following decision variables must be introduced for each EV:

- $t_{\mathrm{in},i}$: starting instant of recharge for the i-th bus;
- $t_{\mathrm{fin},i}$: time instant at which the recharge for the i-th bus ends.

Two additional decision variables, namely ch_i and idle_i, are necessary, which are the time needed for charging the i-th bus and the time between the end of a charging process and the beginning of another one, respectively. They correspond to:

---

[1] Text in this section is © 2023 IEEE. Reproduced from [5], https://doi.org/10.1109/MED51440.2021.9480266; used with permission.

$$\mathrm{ch}_i = t_{fin,i} - t_{in,i} \quad \forall i = 1 \ldots N \tag{6.1}$$

$$\mathrm{idle}_i = t_{in,i} - t_{\mathrm{fin},i-1} \quad \forall i = 1 \ldots N. \tag{6.2}$$

The objective function includes the cost related to the power taken from the grid (that is comprehensive of operational costs, costs for emissions, calculated as a function of an emission factor, and the unit cost for 1 ton of CO_2). That is:

$$\min\ J = \sum_{i=1}^{N} c_{G,i} \tag{6.3}$$

with

$$c_{G,i} = \left\{ \left[\int_{t_{\mathrm{fin},i-1}}^{t_{\mathrm{in},i}} c^E(t)\mathrm{d}t \right] P_{G1,i} + \left[\int_{t_{\mathrm{in},i}}^{t_{\mathrm{fin},i}} c^E(t)\mathrm{d}t \right] P_{G2,i} \right\} \quad i = 1, \ldots, N \tag{6.4}$$

where the power exchange with the grid is defined as $P_{G1,i}$ and $P_{G2,i}$ in the idle and charging intervals, respectively. It is important to note that control variables in the following are considered for idle (identified by subscript 1) and charging (identified by subscript 2) intervals. They are assumed to be constant during each event and this is valid for all control variables described in the following.

Finally, the unit cost $c^E(t)$ considers the energy price as well as the CO_2 emission cost.

Then, the basic state equation of the system, describing the dynamics of the state-of-charge of the storage element $x_S(t)$ can be represented, within a discrete-event setting, as:

$$\begin{aligned} x_{S,i} = x_{S,i-1} &+ \left(\eta_{\mathrm{ch}}\overline{P}^{+}_{S1,i} - \eta_{\mathrm{disch}}\overline{P}^{-}_{S1,i}\right)\mathrm{idle}_i \\ &+ \left(\eta_{\mathrm{ch}}\overline{P}^{+}_{S2,i} - \eta_{\mathrm{disch}}\overline{P}^{-}_{S2,i}\right)\mathrm{ch}_i \quad i = 1, \ldots, N \end{aligned} \tag{6.5}$$

$$\overline{P}_{S1,i} = \overline{P}^{+}_{S1,i} - \overline{P}^{-}_{S1,i} \quad i = 1, \ldots, N \tag{6.6}$$

$$\overline{P}_{S2,i} = \overline{P}^{+}_{S2,i} - \overline{P}^{-}_{S2,i} \quad i = 1, \ldots, N \tag{6.7}$$

$$\overline{P}^{+}_{S1,i}, \overline{P}^{-}_{S1,i}, \overline{P}^{+}_{S2,i}, \overline{P}^{-}_{S2,i} \geq 0 \quad i = 1, \ldots, N \tag{6.8}$$

$$x_{S,0} = x_{S,N} \tag{6.9}$$

where the variables $\overline{P}^{+}_{S1,i}$, $\overline{P}^{-}_{S1,i}$, $\overline{P}^{+}_{S2,i}$, $\overline{P}^{-}_{S2,i}$ are the "positive" and "negative" components of $\overline{P}_{S1,i}$ and $\overline{P}_{S2,i}$ respectively, and η_{ch}, η_{disch} are the charging and discharging efficiencies, respectively.

In Eq. (6.10), the initial state-of-charge $x_{S,0}$ is not known, since the overall system has a periodical behavior. This will be modeled in the following through specific constraints.

The following disjunctive inequalities are introduced to force the mutual exclusiveness of the pair $\overline{P}^{+}_{S1,i}$ and $\overline{P}^{-}_{S1,i}$, and also the pair $\overline{P}^{+}_{S2,i}$ and $\overline{P}^{-}_{S2,i}$:

$$\overline{P}^{+}_{S1,i} \leq \theta_{1,i}B \quad i = 1, \ldots, N \tag{6.11}$$

$$\overline{P}^{-}_{S1,i} \leq \left(1 - \theta_{1,i}\right)B \quad i = 1, \ldots, N \tag{6.12}$$

$$\overline{P}^{+}_{S2,i} \leq \theta_{2,i}B \quad i = 1, \ldots, N \tag{6.13}$$

$$\overline{P}^{-}_{S2,i} \leq \left(1 - \theta_{2,i}\right)B \quad i = 1, \ldots, N \tag{6.14}$$

where B is a big number and $\theta_{1,i}$ and $\theta_{2,i}$ are auxiliary binary variables (equal to one if the battery is discharging and 0 if it is discharging).

Some constraints on the time variables are to be taken into account, namely:

$$t_{fin,i} \leq t_{\text{arr},i} + \text{ch}_{w,i} \quad i = 1, \ldots, N \tag{6.15}$$

$$t_{in,i+1} \geq t_{fin,i} + \varepsilon_1 \quad i = 1, \ldots, N - 1 \tag{6.16}$$

$$t_{\text{in},i} \geq t_{\text{arr},i} + \varepsilon_2 \quad i = 1, \ldots, N \tag{6.17}$$

$$t_{\text{arr}_i} + \text{ch}_i < t_{\text{arr}_{i+1}} + \text{ch}_{i+1} \quad i = 1, \ldots, N. \tag{6.18}$$

In particular, inequality (6.15) imposes that the charging process for the i-th bus must end within the available charging window $\text{ch}_{w,i}$. Constraint (6.16) gives the minimum length of an idle interval, i.e., ε_1. The minimum time difference ε_2 between the arrival time $t_{\text{arr},i}$ and the beginning of the charging process $t_{\text{in},i}$ is introduced in (6.17). Constraints (6.18) are necessary to sequence the charging processes according to the arrival order.

The periodicity of the problem is ensured by introducing the following constraints, where $x_{S,0}$ and $x_{S,N}$ are the states of charge (decision variables) of the storage at the time instants $t_{\text{fin},0}$ and $t_{\text{fin},N}$, respectively:

$$t_{\text{fin},N} = t_{\text{fin},0} + T \tag{6.19}$$

$$x_{S,0} = x_{S,N}. \tag{6.20}$$

Constraints from (6.21) to (6.26) are physical limits on the power flows and the state-of-charge:

$$0 \leq P_{G1,i} \leq P_G^{\text{MAX}} \quad i = 1, \ldots, N \tag{6.21}$$

$$0 \leq P_{G2,i} \leq P_G^{\text{MAX}} \quad i = 1, \ldots, N \tag{6.22}$$

$$0 \leq P_{\text{ch},i} \leq P_{\text{ch}}^{\text{MAX}} \quad i = 1, \ldots, N \tag{6.23}$$

$$P_{s,\min} \leq P_{s1,i} \leq P_{s,\max} \quad i = 1, \ldots, N \tag{6.24}$$

$$P_{s,\min} \leq P_{s2,i} \leq P_{s,\max} \quad i = 1, \ldots, N \tag{6.25}$$

$$x_{s,\min} \leq x_{s,i} \leq x_{s,\max} \quad i = 1, \ldots, N. \tag{6.26}$$

Further constraints related to power balance during idle and charging time intervals. That is:

$$P_{g1,i} + P_{s1,i} = 0 \quad i = 1, \ldots, N \tag{6.27}$$

$$P_{g2,i} + P_{s2,i} = P_{\text{ch},i} \quad i = 1, \ldots, N. \tag{6.28}$$

Finally, it is necessary to introduce a constraint related to the satisfaction of the energy demand for each vehicle. That is:

$$\text{EC}_i = \eta_i P_{\text{ch},i} \text{ch}_i \quad i = 1, \ldots, N \tag{6.29}$$

where η_i is the efficiency of the EBs' battery and represents energy consumption.

It is important to note that, in the case of EBs, EC_i corresponds to the bus energy consumption during the daily trips. This value can be calculated through the equations presented in Chap. 2.

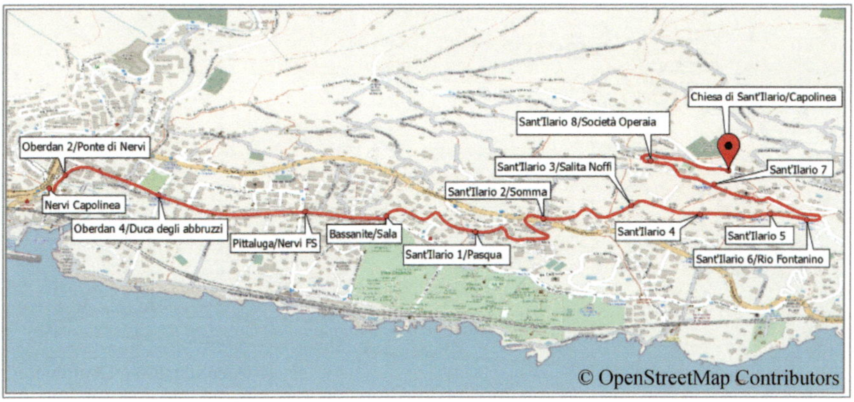

Fig. 6.3 Line 516 path [Base map from OpenStreetMap (https://www.openstreetmap.org/copyright); Available under the Open Database License (https://opendatacommons.org/). Red route line and labels are generated with QGIS (http://www.qgis.org).]

6.2.1 *Application to a Case Study*[2]

This section applies the proposed approach to a specific case study. A base scenario is first considered, where the number of buses to be charged is five ($N = 5$).

The solve the nonlinear optimization problem previously defined, the software tool LINGO™ has been used.

The buses considered in the study are vehicles that work on lines 516 and 517 of the bus company operating in Genoa (AMT). In particular, on the software QGIS, a project for each route has been realized in order to calculate the mileage, the duration of the runs has been calculated based on the information reported by some data collected on the buses and, finally, Google Earth Pro has allowed calculating the overall height difference of each route. The height difference is calculated since it affects fuel consumption and the possibility of regenerative braking.

Line 516 takes approximately 22 min to complete a full ride. The outward and return routes (Fig. 6.3) are equal, and the overall length is 6 km with 24 stops and a total height difference of about 280 m.

Note that the bus must go to the Nervi terminus to start the assigned run at the beginning of the service and therefore travels about 8.9 km from the depot.

In line 517, a complete round trip takes about 22 min. Unlike line 516, the two round-trip routes (Figs. 6.4 and 6.5) are very different: on the outward journey, the bus reaches the Nervi cemetery and makes 22 stops, covering a distance of about 4.5 km, while on the return journey, it covers a distance of about 1.8 km divided into six stops. The outward journey covers a total height difference of about 127 m, while the return journey covers about 65 m. Also, in this case, at the beginning of the service, the bus must go to the Nervi terminus to start the assigned run.

---

[2] Text in this section is © 2023 IEEE. Reproduced from [5], https://doi.org/10.1109/MED51440.2021.9480266; used with permission.

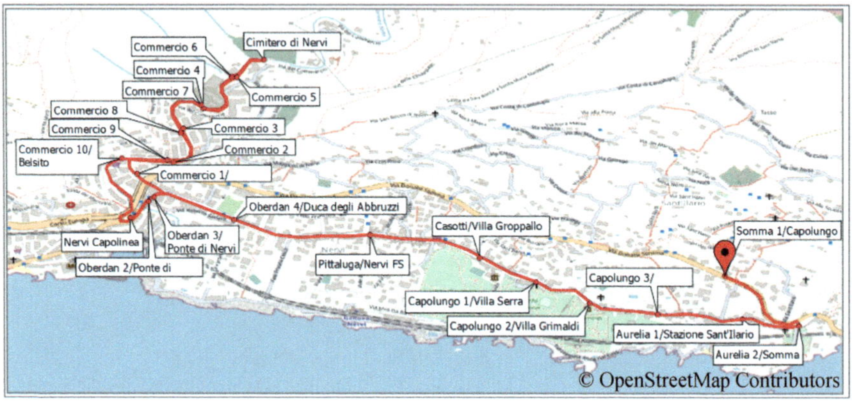

Fig. 6.4 Outward path of line 517 [Base map from OpenStreetMap (https://www.openstreetmap.org/copyright); Available under the Open Database License (https://opendatacommons.org/). Red route line and labels are generated with QGIS (http://www.qgis.org).]

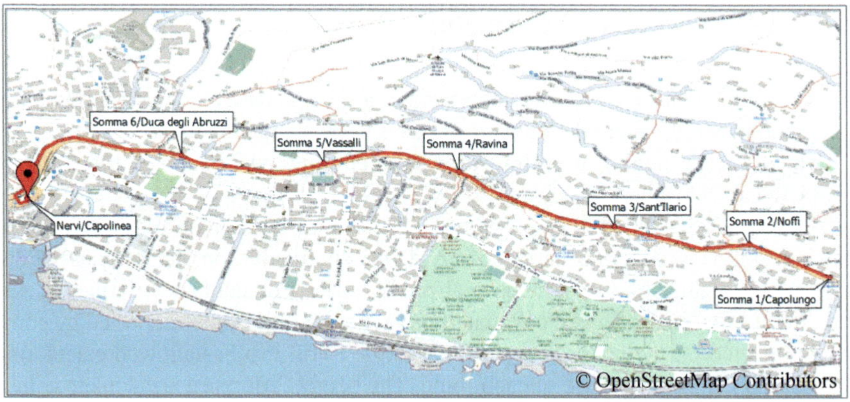

Fig. 6.5 Backward path of line 517 [Base map from OpenStreetMap (https://www.openstreetmap.org/copyright); Available under the Open Database License (https://opendatacommons.org/). Red route line and labels are generated with QGIS (http://www.qgis.org).]

The data considered for the case study are presented in Tables 6.1 and 6.2. The values relevant to the calculation of the energy consumption EC_i, for each of the considered paths, have been determined using the software tool QGIS.

In particular, note that the first two buses are assigned to line 516 and the other three buses are assigned to line 517.

To calculate the integral functions in (6.4), the cost function has been created via a sixth-order polynomial approximation of a stairs function, in the form:

$$p(x) = a_6x^6 + a_5x^5 + a_4x^4 + a_3x^3 + a_2x^2 + a_1x^1 + a_0.$$

The coefficients are reported in Table 6.3.

Table 6.1 Data relevant to buses

i	$t_{arr,i}$ [hh:mm:ss]	$ch_{w,i}$ [hh:mm:ss]	EC_i [kWh]
1	12:00:00	3:12:00	65.55
2	12:43:00	6:22:00	76.24
3	16:58:00	13:22:00	75.86
4	17:12:00	14:00:00	52.29
5	18:49:00	17:58:00	59.42

Table 6.2 System parameters

Parameter	Value
CAP	180 [kWh]
$\varepsilon 1$	0.1 [h]
$\varepsilon 2$	0.1 [h]
$x_{\min}$	0.1 [–]
$x_{\max}$	0.9 [–]
$P_{g,\max}$	20 [kW]
$P_{ch,\max}$	30 [kW]
$P_{s,\min}$	– 30 [kW]
$P_{s,\max}$	30 [kW]

Table 6.3 Polynomial function coefficient

Parameter	Value	Parameter	Value
a_6	– 1.18E – 09	a_2	– 2.37E – 03
a_5	1.70E – 07	a_1	8.35E – 03
a_4	– 9.10E – 06	a_0	2.04E – 01
a_3	2.22E – 04		

Note that this curve presents a lower cost around midday and a higher cost around midnight

6.2.1.1 Base Scenario

In this scenario, the optimization problem results show that five buses can be recharged with one charging station. Therefore, it is proven that considerable economic benefits can be achieved by saving on the initial investment for the purchase and installation of the charging stations.

According to the results obtained, the total charging cost is equal to 70.87€, and the system starts the process cycle with a storage state-of-charge of about 0.5.

Table 6.4 shows the results obtained using the LINGO software.

Figures 6.6, 6.7 and 6.8 show the power profiles.

As it is possible to see from the figures, the maximum power available from the grid is purchased for four buses out of five. The storage element remains almost

Table 6.4 Results with $N = 5$

BUS i	$t_{\text{fin},i-1}$	$t_{\text{in},i}$	$P_{g1,i}$	$P_{s1,i}$	$P_{g2,i}$	$P_{s2,i}$	$P_{\text{ch},i}$
BUS 1	11.51	12.10	20.00	– 20.00	20.00	2.24	22.24
BUS 2	15.20	15.30	20.00	– 20.00	20.00	1.28	21.28
BUS 3	19.07	19.17	0.00	0.00	20.00	0.00	20.00
BUS 4	23.16	27.45	0.00	0.00	14.68	0.00	14.68
BUS 5	31.20	32.39	0.00	0.00	20.00	0.00	20.00

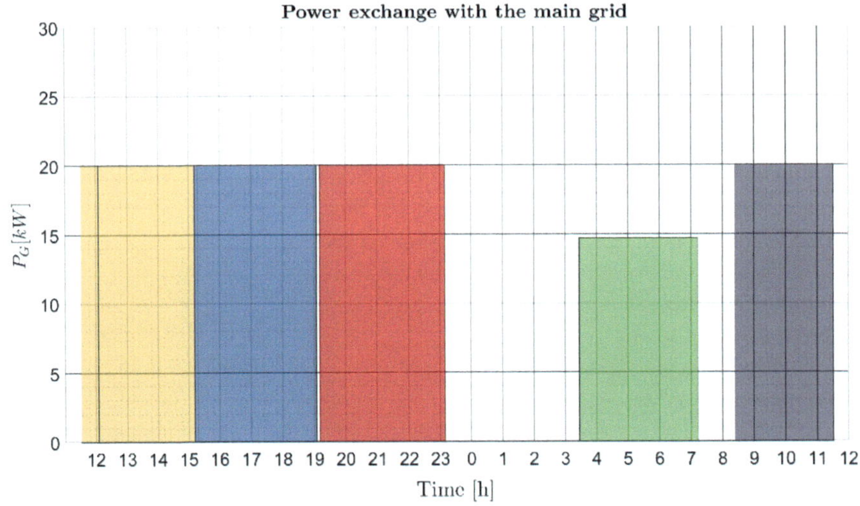

Fig. 6.6 Power purchased from the grid

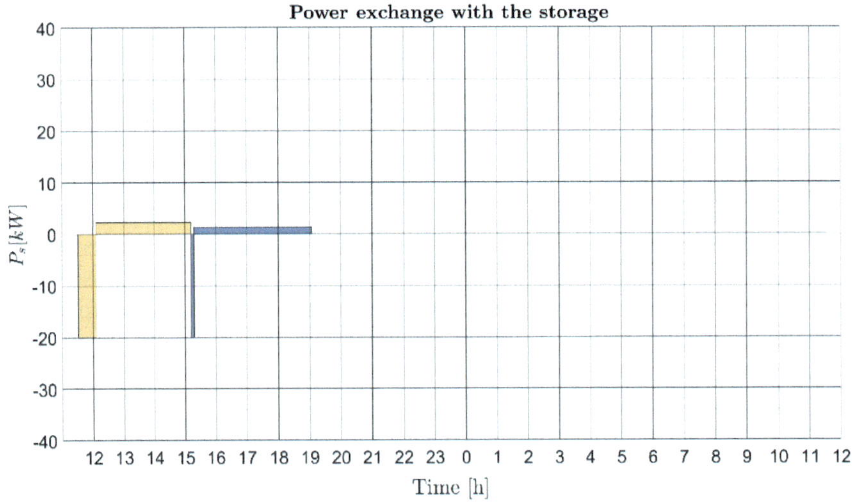

Fig. 6.7 Power exchanged with the storage

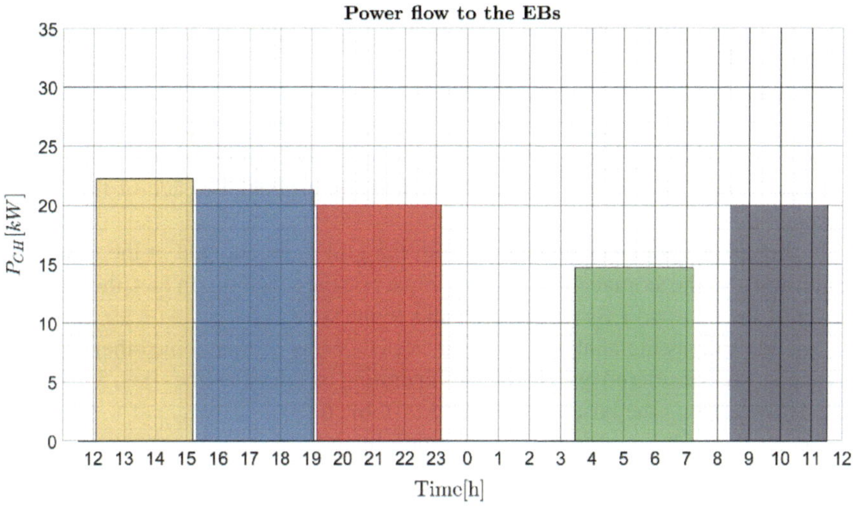

Fig. 6.8 Charging power

unused but for the first time intervals in which the power is used to charge the buses. There is a long idle time between the third and the fourth vehicle because of the time constraints and the cost functions, which decreases in the night.

6.2.1.2 Sensitivity Analysis

The sensitivity analysis consists of evaluating the effects on the optimization problem results by varying some uncertain parameters. Therefore, an analysis of this type aims to improve the decision-making process by assessing the robustness of the decision made and highlighting factors whose value should be better estimated and those that should be kept under strict control during execution.

In the proposed optimization problem, the parameter subject to greater uncertainty is the energy demand of each bus. This parameter depends on several factors such as:

- the bus typology;
- the driver's driving style;
- the weather conditions and the use of auxiliary devices (air conditioning);
- the number of passengers on board;
- traffic conditions;
- the number of start and stops during the service.

Therefore, it was decided to perform a sensitivity analysis to observe the system's response if the energy demand varies. In particular, we considered an increase of 20 kWh in the energy requests concerning the values reported in Table 6.1. Then, we tried to add other vehicles to determine if the system could also include other charging processes.

The results show that by increasing the bus energy demand by 20 kWh, the overall recharge cost turns out to be 94.30€, and the system starts the process cycle with a storage state-of-charge level of 0.55. Table 6.5 reports the results obtained.

Figures 6.9, 6.10 and 6.11 show the absorption profiles for each of the powers considered.

In this case, the power purchased from the grid is maintained at the maximum value, and also the storage gains a larger impact on the schedule. In fact, the charging power of the first two EBs is higher, thanks to the storage.

Another interesting analysis has been carried out by adding some other buses. In particular, we considered two more buses operating on the same lines (i.e., 516 – 517). The data regarding the buses are reported in Table 6.6.

According to the presented formalization, they have been inserted considering the arrival order, and then in the final order they will hold positions three and seven, respectively. The results obtained with the additional buses are reported in Table 6.7.

Table 6.5 Results considering an energy request increased by 20 kWh

BUS i	$t_{\text{fin},i-1}$	$t_{\text{in},i}$	$P_{g1,i}$	$P_{s1,i}$	$P_{g2,i}$	$P_{s2,i}$	$P_{\text{ch},i}$
BUS 1	10.25	12.10	20.00	– 20.00	20.00	9.03	29.03
BUS 2	15.20	15.30	20.00	– 20.00	20.00	6.86	26.86
BUS 3	19.07	19.17	20.00	– 20.00	20.00	0.00	20.00
BUS 4	24.21	25.17	0.00	0.00	20.00	– 4.15	15.85
BUS 5	29.98	30.08	20.00	– 20.00	20.00	0.00	20.00

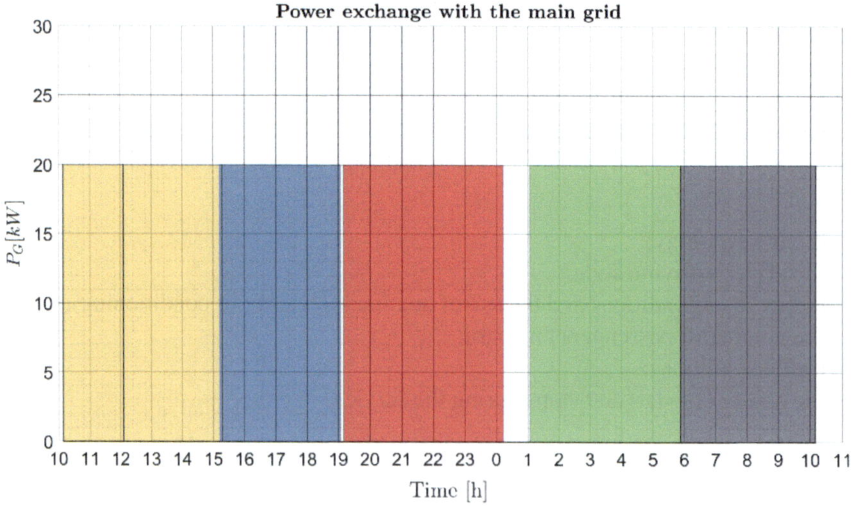

Fig. 6.9 Power purchased from the grid (EC + 20 kWh)

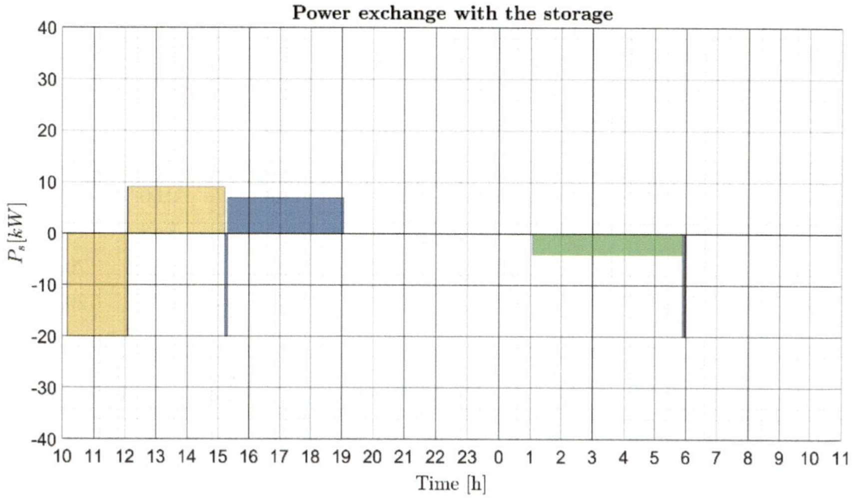

Fig. 6.10 Power exchanged with the storage (EC + 20 kWh)

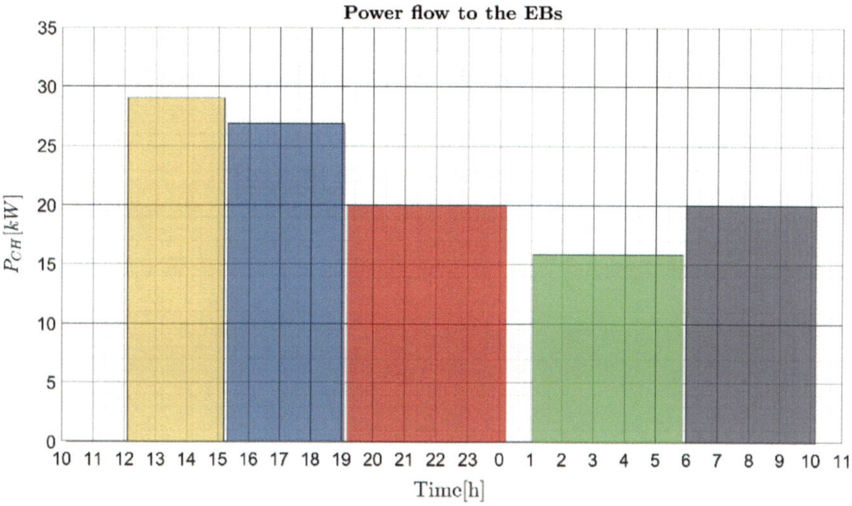

Fig. 6.11 Charging power (EC + 20 kWh)

Table 6.6 Data relevant to the additional buses

i	$t_{\text{arr},i}$ [hh:mm:ss]	$\text{ch}_{w,i}$ [hh:mm:ss]	EC_i [kWh]
1	14:06:00	7:30:00	55.00
2	19:00:00	18:00:00	58.30

Table 6.7 Results considering two additional buses

BUS i	$t_{\text{fin},i-1}$	$t_{\text{in},i}$	$P_{g1,i}$	$P_{s1,i}$	$P_{g2,i}$	$P_{s2,i}$	$P_{\text{ch},i}$
BUS 1	12.00	12.10	20.00	– 20.00	20.00	8.79	28.79
BUS 2	14.49	14.59	20.00	– 20.00	20.00	0.00	20.00
BUS 3	18.61	18.71	20.00	– 20.00	20.00	0.00	20.00
BUS 4	21.60	21.70	0.00	0.00	20.00	0.00	20.00
BUS 5	25.69	25.86	0.00	0.00	18.57	– 3.88	14.68
BUS 6	29.61	29.71	20.00	– 20.00	20.00	0.00	20.00
BUS 7	32.83	32.93	20.00	– 20.00	20.00	0.00	20.00

Figures 6.12, 6.13 and 6.14 are the absorption profiles for each of the powers considered.

In this case, the use of storage is limited because of the limits on time and maximum power from the grid. It is possible to note that the storage is discharged only in the first charging process, and then it is charged during the remaining idle intervals. As it is possible to see, the 24 h horizon is almost filled. An instance with eight buses has been carried out, but as expected, the system could not satisfy the request. A possible solution could be increasing the maximum power available from the grid. In the following (Table 6.8), we summarize the overall cost obtained in the different instances.

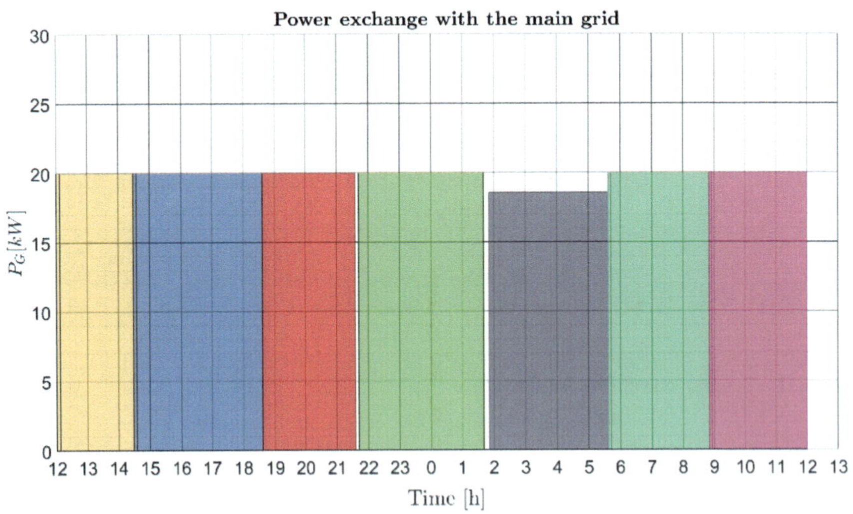

Fig. 6.12 Power purchased from the grid ($N = 7$)

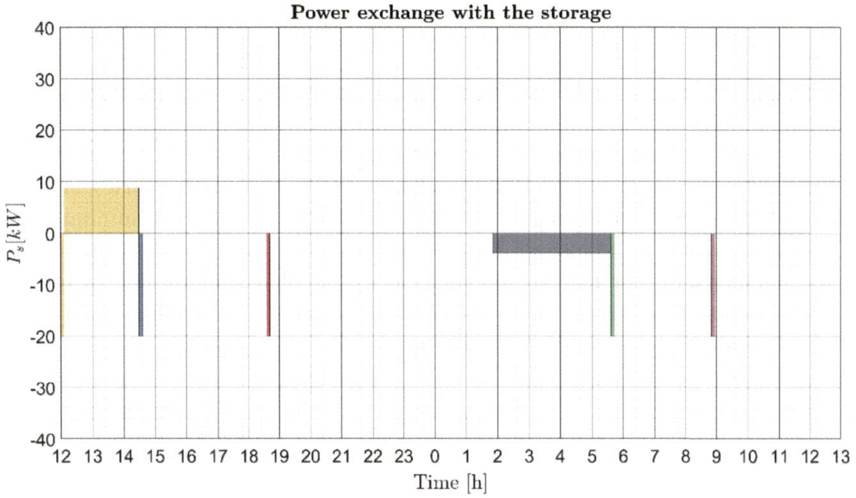

Fig. 6.13 Power exchanged with the storage ($N = 7$)

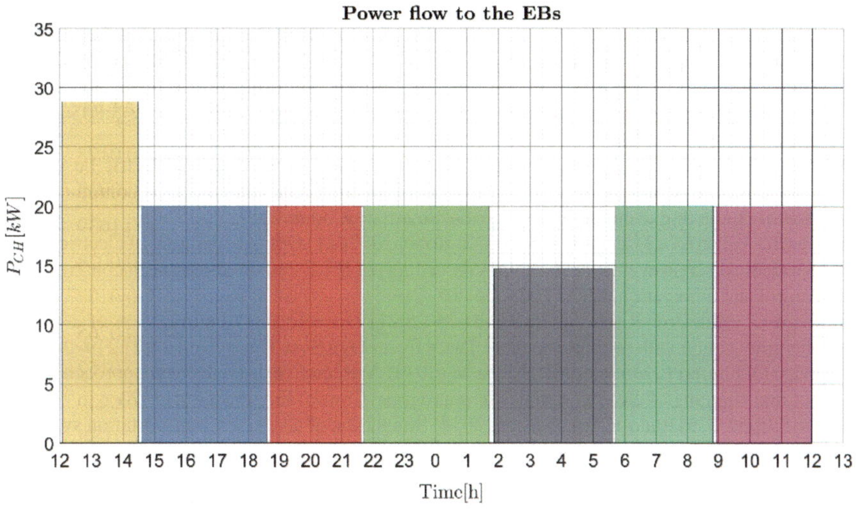

Fig. 6.14 Charging power ($N = 7$)

Table 6.8 Costs comparison

Instance	Cost [€]	Run time [s]
$N = 5$	70.87	1
$N = 5$ (+ 20 kWh)	94.31	1
$N = 6$	83.36	1
$N = 7$	96.16	1

6.2.2 Conclusion

In this chapter, a new model based on a periodic discrete-event approach for the optimal schedule of the charging of EBs has been presented. With respect to the existing literature, in this paper, we propose a new periodic discrete-event approach. The considered system is composed of a charging station, N electric buses, storage, and a connection with the main grid. The objective function includes the minimization of the overall costs. A first scenario based on real data has been considered in the case study, and then a further analysis has been carried out increasing the energy request for each bus and then the number of buses. With the considered configuration of the system, a limit of eight buses has been determined.

References

1. Deng R, Liu Y, Chen W, Liang H (2021) A survey on electric buses—energy storage, power management, and charging scheduling. IEEE Trans Intell Transp Syst 22(1):9–22. https://doi.org/10.1109/TITS.2019.2956807
2. Jahic A, Plenz M, Eskander M, Schulz D (2021) Route scheduling for centralized electric bus depots. IEEE Open J Intell Transp Syst: 1. https://doi.org/10.1109/OJITS.2021.3096115
3. Zhou GJ, Xie DF, Zhao XM, Lu C (2020) Collaborative optimization of vehicle and charging scheduling for a bus fleet mixed with electric and traditional buses. IEEE Access 8:8056–8072. https://doi.org/10.1109/ACCESS.2020.2964391
4. Zhang S, Leung KC (2019) Joint optimal allocation and scheduling for electric buses with vehicle-to-grid regulation service. In: 2019 IEEE international conference on communication control computing technology smart grids, SmartGridComm 2019
5. Casella V, Ferro G, Minciardi R, Parodi L, Robba M (2021) Optimal charging of electric buses: a periodic discrete event approach. In: Proceedings of IEEE 29th Mediterranean conference on control and automation, MED 2021, pp 208–213
6. Ferro G, Minciardi R, Parodi L, Robba M (2021) Discrete event optimization of a vehicle charging station with multiple sockets. Discr Event Dyn Syst Theor Appl 31(2):219–249
7. Ferro G, Laureri F, Minciardi R, Robba M (2019) A predictive discrete event approach for the optimal charging of electric vehicles in microgrids. Control Eng Pract 86:11–23
8. De Filippo G, Marano V, Sioshansi R (2014) Simulation of an electric transportation system at The Ohio State University. Appl Energy 113:1686–1691. https://doi.org/10.1016/j.apenergy.2013.09.011
9. Deilami S, Masoum AS, Moses PS, Masoum MAS (2011) Real-time coordination of plug-in electric vehicle charging in smart grids to minimize power losses and improve voltage profile. IEEE Trans Smart Grid 2(3). https://doi.org/10.1109/TSG.2011.2159816
10. Lajunen A (2014) Energy consumption and cost-benefit analysis of hybrid and electric city buses. Transp Res Part C Emerg Technol 38:1–15. https://doi.org/10.1016/j.trc.2013.10.008
11. Liu Z, Song Z, He Y (2018) Planning of fast-charging stations for a battery electric bus system under energy consumption uncertainty. Transp Res Rec 2672(8):96–107. https://doi.org/10.1177/0361198118772953
12. Lin Y, Zhang K, Shen Z-JM, Ye B, Miao L (2019) Multistage large-scale charging station planning for electric buses considering transportation network and power grid. Transp Res Part C Emerg Technol 107:423–443. https://doi.org/10.1016/j.trc.2019.08.009
13. Ferro G, Minciardi R, Parodi L, Robba M (2021) Optimal planning of charging stations in coupled transportation and power networks based on user equilibrium conditions. IEEE Trans Autom Sci Eng (in Press)

14. Xylia M, Leduc S, Patrizio P, Kraxner F, Silveira S (2017) Locating charging infrastructure for electric buses in Stockholm. Transp Res Part C Emerg Technol 78:183–200. https://doi.org/10.1016/j.trc.2017.03.005

Chapter 7
Planning for Electric Vehicles: A Stochastic User Equilibrium Approach for Joint Traffic Assignment and Energy Demand Assignment

7.1 Introduction

In this chapter, we focus our attention on the formalization of the joint traffic and energy demand assignment conditions over a traffic network in the presence of electric vehicles. Specifically, we will use the so-called stochastic user equilibrium (SUE) conditions to define the steady-state behavior of a generic traffic network with some arcs equipped with EV charging stations. Generally, this kind of analysis is carried out in the deterministic setting, assuming that travelers choose the least cost path from their origin to their destination. Although this assumption may not seem unreasonable, it presumes that all travelers have perfect information regarding the costs (e.g., travel time and recharging time and cost) over the entire network, that they make consistently correct decisions, and that they all behave identically [2]. From a practical point of view, it is evident that these assumptions are not too realistic. The technique used in this chapter, based on the SUE conditions, allows avoiding unrealistic situations in which, as an example, paths having similar costs are not assigned similar flows.

The SUE model was first proposed in [3] and is a typical method to solve the traffic assignment problem since it provides a good equilibrium traffic flow pattern by considering travelers' imperfect perceptions of travel cost. SUE conditions are affected by the user's route choice model [4]. Many studies are presented in current literature like [5]. The authors studied the impact of choice function and parameters estimation performance (log-likelihood values) of route choice models applied to a real-world urban network. An interesting study on the different types of equilibria is

G. Ferro et al., *Optimization of Electric-Vehicle Charging*, Advances in Industrial Control, https://doi.org/10.1007/978-3-031-61917-5_7

carried out in [6], in which the differences between deterministic and stochastic equilibrium are analyzed using supernetworks. SUE conditions are used as constraints to define the traffic demand in the case of the location of EV charging stations. As an example, in [7] the SUE conditions are used to capture the maximum EV path flow on a network; in [8] an equilibrium-based modeling framework is considered for locating plug-in charging facilities, which includes transportation and power networks and maximizes social welfare. In [9], a study is presented that considers electricity prices and road pricing as instruments to improve the management of transportation and power networks.

This chapter aims to formalize and test the SUE conditions for mixed flows of vehicles (traditional and electric) with different travel costs in the presence of arcs equipped with charging stations. This gives rise to a system of nonlinear equations whose solution corresponds to a feasibility problem.

It is important to note that the model in this chapter has its basis in the work proposed in [10].

The following section presents the traffic network model with different costs for paths with and without EV charging stations equipped. Then, Sect. 7.3 describes the SUE conditions under two different user's path choice functions, namely the multinomial logit and the multinomial C-logit. Finally, in Sect. 7.4, the SUE conditions defined in the previous section will be simulated over a modified Nguyen–Dupuis network.

7.2 Traffic Network Modeling

We define: A as the set of directed links of a traffic network, S as the origin set, and D as the destination set. Each origin–destination pair (s, d), $s \in S$ and $d \in D$, has a specified traffic demand that is characterized by two different terms:

- q^1_{sd} the demand of traditional vehicles or EV without charging necessity;
- q^2_{sd} the demand of EV with charging necessity.

In order to consider the second kind of demand we define as R the subset of A in which the links have a charging station. To this end, we augment the original network such that a link belonging to R is split in two links, namely a and $\tilde{a}$ i.e., the link when covered by a traffic that do not need a recharge and the links where EVs need to be charged. In this way, we introduce $\tilde{R}$ as the set of "additional" links used to perform this separation. Therefore, we create an augmentation of the original graph with arcs $A \cup \tilde{R}$.

Now for any (s, d) pair, we can define the set of paths:

- P^1_{sd} (with cardinality K^1_{sd}) are the paths for vehicles that do not need to recharge during the trip;
- P^2_{sd} (with cardinality K^2_{sd}) are the paths for EV that need to stop at the charging station.

To begin with the mathematical modeling, we introduce the following coefficients $\delta_{a,k}^{sd}$ and $\xi_{a,k}^{sd}$ defined as:

$$\delta_{a,k}^{sd} = \begin{cases} 1 & \text{if link } a \in \text{ the } k\text{-th path in } P_{sd}^1 \\ 0 & \text{otherwise} \end{cases} \quad a \in A, (s,d) \in O/D, k = 1, \ldots, K_{sd}^1 \tag{7.1}$$

$$\xi_{a,k}^{sd} = \begin{cases} 1 & \text{if link } a \in \text{ to the } k\text{-th path in } P_{sd}^2 \\ 0 & \text{otherwise} \end{cases} \quad a \in \text{AU}\tilde{\text{R}}, (s,d) \in O/D, k = 1, \ldots, K_{sd}^2. \tag{7.2}$$

By hypothesis, EVs never require more than one charging stop during the trip, and thus indexes $\xi_{a,k}^{sd}$ must fulfill the condition:

$$\sum_{a \in \tilde{R}} \xi_{a,k}^{sd} = 1 \quad (s,d) \in O/D, \quad k = 1, \ldots, K_{sd}^2. \tag{7.3}$$

At this time, we will focus on the flows on each path. f_k^{sd} is the flow of vehicles which do not have any recharge in path k. To make the model consistent, the sum of the flows on each path must be equal to the traffic demand, that is:

$$q_{sd}^1 = \sum_{k=1}^{K_{sd}^1} f_k^{sd} \quad \forall (s,d) \in O/D. \tag{7.4}$$

In the same way, g_k^{sd} are the flows relevant to vehicles that need a recharge in path k. Analogously to (4) we state the following relation:

$$q_{sd}^2 = \sum_{k=1}^{K_{sd}^2} g_k^{sd} \quad \forall (s,d) \in O/D. \tag{7.5}$$

Thanks to the previous definitions, it is possible to define the flows on each link $a \in Aa \in A$ as the sum of the flows f_k^{sd} in paths P_{sd}^1 that include link $a \in A$, plus the flow g_k^{sd} in P_{sd}^2 with link $a \in A$. That is:

$$x_a = \sum_{(s,d) \in O/D} \sum_{k=1}^{K_{sd}^1} f_k^{sd} \delta_{a,k}^{sd} + \sum_{(s,d) \in O/D} \sum_{k=1}^{K_{sd}^2} g_k^{sd} \xi_{a,k}^{sd} \quad a \in A. \tag{7.6}$$

As regards a link $a \in \tilde{R}$, since there is no flow f_k^{sd}, we have:

$$x_a = \sum_{(s,d)\in O/D} \sum_{k=1}^{K_{sd}^2} g_k^{sd} \xi_{a,k}^{sd} \quad a \in \tilde{R}. \tag{7.7}$$

7.2.1 The Cost Expression for Each Link

According to [11], the travel time in the link $a \in A$ is assumed to be obtained by the following relationship:

$$t_a(x_a) = t_a^0 \left[1 + \alpha_a \left(\frac{x_a}{\mathrm{CAP}_a} \right)^4 \right] \quad a \in A \tag{7.8}$$

where:

- t_a^0 represents the link's free-flow travel time [h];
- α_a is an adimensional parameter tuned based on real data;
- CAP_a [#veh/h] is the link's maximum flow.

We model the energy demand of EVs as a deterministic parameter. We assume that the recharging service demand is the same, for the sake of simplicity, for each vehicle class independent of the O/D path.

Then, we can write:

$$y_a = \sum_{(s,d)\in O/D} \sum_{k=1}^{K_{sd}^2} g_k^{sd} \xi_{a,k}^{sd} \overline{ER} \quad a \in \tilde{R} \tag{7.9}$$

where $\overline{ER}$ is in [kWh], g_k^{sd} is in [veh/h], and y_a [kWh/h] is the energy request per unit of time in the considered link/station. Therefore, the generalized cost $c_{tot,a}(x_a, y_a)$ [h] for charging links is given by:

$$c_{\mathrm{tot},a}(x_a, y_a) = c_{1,a}(x_a) + c_{2,a}(y_a) \quad a \in \tilde{R} \tag{7.10}$$

where

$$c_{1,a}(x_a) = t_a(x_a) \quad a \in \tilde{R} \tag{7.11}$$

$$c_{2,a}(y_a) = ts_a^0 + \tilde{c}_{2,a}(y_a) + \omega p_a \overline{ER} \quad a \in \tilde{R} \tag{7.12}$$

$$\tilde{c}_{2,a}(y_a) = \beta_a \left(\frac{y_a}{\text{CAPs}_a - y_a} \right) \quad a \in \tilde{R} \tag{7.13}$$

where:

- ts_a^0 is the pure service time [h];
- β_a is a parameter characteristic of the charging station;
- CAPs_a is the charging station capacity in link a [# of services/h];
- p_a is the power unit price [€/kWh] in link a;
- ω is a conversion parameter [h/€].

Note that we have used the Davidson [12] formula (7.13), to express an estimate of the waiting time before service at the considered station, as a monotone function of y_a.

7.2.2 Expressing the Cost for the Various Paths[1]

The total cost expressed in hours, associated with each vehicle following the k-th path from s to d belonging to P_{sd}^1 corresponds to the overall travel time over the links included in that path. Namely, it can be defined as:

$$c_{\text{path},k}^{sd} = \sum_{a \in A} t_a(x_a)\delta_{a,k}^{sd} \quad k = 1, \ldots, K_{sd}^1 \tag{7.14}$$

where $t_a(x_a)$ is defined by (7.8).

Instead, the cost relevant to vehicles traveling over paths belonging to P_{sd}^2 can be determined by the sum of two main terms:

- the overall travel time over the "ordinary" (i.e., those belonging to A) links included in that path;
- the cost $c_{\text{tot},a}(x_a, y_a)$ incurred whenever a recharge of a vehicle in the link $a \in \tilde{R}$ takes place.

That is:

$$\tilde{c}_{\text{path},k}^{sd} = \sum_{a \in A} t_a(x_a)\xi_{a,k}^{sd} + \sum_{a \in \tilde{R}} c_{\text{tot},a}(x_a, y_a)\xi_{a,k}^{sd} \quad k = 1, \ldots, K_{sd}^2. \tag{7.15}$$

After the definition of the cost functions on each path, it is possible to state the SUE conditions.

---

[1] Text in this section is © 2020 IEEE. Reproduced from [10], https://doi.org/10.1016/j.ifacol.2020.12.157; used with permission.

7.3 Stochastic User Equilibrium Conditions

The main assumption behind the deterministic assignment model is the perfect knowledge of network conditions and therefore the travel cost on each route. Since this assumption is unrealistic in practice, in [13] a first distinction between the drivers perceived and the actual travel cost has been carried out. This is the basis of the SUE principle: no traveler can improve its *perceived* cost by individually changing his/her route. The route choice model is probabilistic and is derived by assuming a deterministic component, $\overline{c}^{sd}_{\text{path},k}$ plus a random error, $\theta^{sd}_{\text{path},k}$, namely:

$$c^{sd}_{\text{path},k} = \overline{c}^{sd}_{\text{path},k} + \theta^{sd}_{\text{path},k} \quad (s,d) \in O/D, \quad k = K^1_{sd} \cup K^2_{sd}. \tag{7.16}$$

The random component of the utility satisfies $E\left[\theta^{sd}_{\text{path},k}\right] = 0$, meaning that $E\left[c^{sd}_{\text{path},k}\right] = \overline{c}^{sd}_{\text{path},k}$.

In this case, the equilibrium model depends both on the congestion effect and on the stochastic effects.

Given the O-D trip rates q^1_{sd} and q^2_{sd}, the following equations characterize the stochastic equilibrium conditions:

$$f^{sd}_k = Q^{sd}_k q^1_{sd} \quad \forall (s,d) \in O/D, \quad k \in K^1_{sd} \tag{7.17}$$

$$g^{sd}_k = \hat{Q}^{sd}_k q^2_{sd} \quad \forall (s,d) \in O/D, \quad k \in K^2_{sd} \tag{7.18}$$

where $\hat{Q}^{sd}_k$ and Q^{sd}_k are the probabilities that route k between s and d is chosen, for the two categories of routes. The probability that a given route is chosen is the probability that its travel time is perceived to be the lowest of all the alternative routes. In the following sections, two different probability functions will be considered, namely the multinomial logit and the multinomial C-logit.

7.4 Logit Model

The multinomial logit model [13], when it is used to generate probabilities for paths K^1_{sd} and K^2_{sd}, provides:

$$Q^{sd}_k = \frac{e^{-c^{sd}_{\text{path},k}}}{\sum_{k \in K^1_{sd}} e^{-c^{sd}_{\text{path},k}}} \quad \forall (s,d) \in O/D, k \in K^1_{sd} \tag{7.19}$$

$$\hat{Q}^{sd}_k = \frac{e^{-\hat{c}^{sd}_{\text{path},k}}}{\sum_{k \in K^2_{sd}} e^{-\hat{c}^{sd}_{\text{path},k}}} \quad \forall (s,d) \in O/D, k \in K^2_{sd}. \tag{7.20}$$

The choice model constraints (7.19) and (7.20) are used to represent the drivers' routing choices. In this model, the main issues are the flow-dependent (overall) travel times, which consider the network flow congestion effects and the waiting time at the charging station.

It has been pointed out in [14] that a noticeable drawback of the multinomial logit model is that it cannot account for similarities among different alternatives (the paths, in our case), i.e., overlapping among different routes. For this reason, it may give rise to unrealistic route choice probabilities. The next section will adopt a modified version of the stochastic model that tries to overcome this drawback.

7.5 C-Logit Model

To overcome the problems of the multinomial logit route choice model, the C-logit model [14] uses the *commonality factor* $cf^{sd}_{path,k}$ that considers also overlapping paths:

$$Q^{sd}_k = \frac{e^{-\left(c^{sd}_{\text{path},k}+cf^{sd}_{\text{path},k}\right)}}{\sum_{k\in K^1_{sd}} e^{-\left(c^{sd}_{\text{path},k}+cf^{sd}_{\text{path},k}\right)}} \quad \forall(s,d)\in O/D, k\in K^1_{sd} \tag{7.21}$$

$$\hat{Q}^{sd}_k = \frac{e^{-\left(\hat{c}^{sd}_{\text{path},k}+cf^{sd}_{\text{path},k}\right)}}{\sum_{k\in K^2_{sd}} e^{-\left(\hat{c}^{sd}_{\text{path},k}+cf^{sd}_{\text{path},k}\right)}} \quad \forall(s,d)\in O/D, k\in K^2_{sd}. \tag{7.22}$$

The commonality factor is defined as:

$$cf^{sd}_{\text{path},k} = \beta \ln\left[\sum_{l\in K^1_{sd}}\left(\frac{L_{lk}}{\sqrt{L_l}\sqrt{L_k}}\right)\right] \quad \forall(s,d)\in O/D, k\in K^1_{sd} \tag{7.23}$$

where L_{lk} is the length of links common to routes l and k, L_l and L_k are the overall length of routes l and k, respectively, and β is a parameter having the dimension of a cost (a similar relationship can be written for paths belonging to the set P^2_{sd}).

It has been shown by [14] that the C-logit model captures the overlapping effect among the different paths via commonality factors. Therefore, it can capture a more realistic traffic flow distribution pattern.

7.6 Case Study

The proposed steady-state study involves the solution of two nonlinear systems of equations, namely (7.4)–(7.20) for the logit model and (7.4)–(7.18) and (7.21)–(7.23) for the C-logit model. The considered network is a modified version of the popular Nguyen–Dupuis network [15] depicted in Fig. 7.1. In the considered case study, there are 13 nodes, 23 links of which 4 are related to recharging stations.

The data for each arc, traditional and electric ones, are reported in Tables 7.1 and 7.2.

Since there are 2 origin nodes (1 and 4) and 2 destination nodes (2 and 3) in the test network we have 4 different classes of non-electric paths, namely P^1_{12}, P^1_{13}, P^1_{42}, and P^1_{43}. The following Tables 7.3, 7.4, 7.5 and 7.6 enumerates the paths (link sequences) for each class.

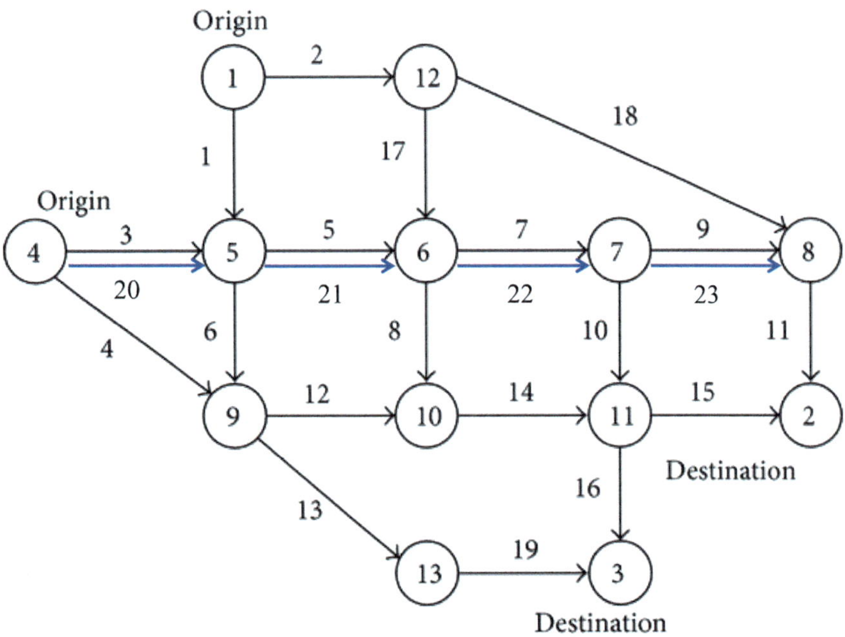

Fig. 7.1 Case study network

Table 7.1 Data for non-electric links

Arc #	t_a^0	CAP	α_a
1	0.1	1000	32
2	0.17	1000	32
3	0.08	1000	32
4	0.15	750	32
5	0.25	400	32
6	0.25	500	32
7	0.08	500	32
8	0.2	750	32
9	0.25	400	32
10	0.25	500	32
11	0.1	1000	32
12	0.17	1000	32
13	0.08	1000	32
14	0.15	750	32
15	0.25	400	32
16	0.25	500	32
17	0.08	500	32
18	0.2	750	32
19	0.25	400	32

Table 7.2 Data for electric links

Arc #	t_a^0	CAP	α_a	$ts_a^{sd,0}$	$CAPs_a$	β_a	$p_{r,a}$
20	0.1	1000	32	0.5	45	2	15
21	0.17	1000	32	0.3	55	2	15
22	0.08	1000	32	0.3	60	2	15
23	0.15	750	32	0.3	60	2	15

Table 7.3 Paths for P_{12}^1

#Path	P_{12}^1 links				
1	1	5	7	9	11
2	2	18	11		
3	1	5	7	10	15
4	1	5	8	14	15
5	1	6	12	14	15
6	2	17	7	9	11
7	2	17	7	10	15
8	2	17	8	14	15

Table 7.4 Paths for P_{13}^{1}

#Path	P_{13}^{1} links				
1	1	5	8	14	16
2	2	17	8	14	16
3	1	5	7	10	16
4	1	6	12	14	16
5	1	6	13	19	
6	2	17	7	10	16

Table 7.5 Paths for P_{42}^{1}

#Path	P_{42}^{1} links				
1	3	5	7	9	11
2	4	12	14	15	
3	3	5	7	10	15
4	3	5	8	14	15
5	3	6	12	14	15

Table 7.6 Paths for P_{43}^{1}

#Path	P_{43}^{1} links				
1	3	5	8	14	16
2	4	13	19		
3	3	5	7	10	16
4	3	6	12	14	16
5	3	6	13	19	
6	4	12	12	16	

For each class of paths, the link to path matrices are:

$$\delta^{1,2} = \begin{bmatrix} 1&0&1&1&1&0&0&0\\ 0&1&0&0&0&1&1&1\\ 0&0&0&0&0&0&0&0\\ 0&0&0&0&0&0&0&0\\ 1&0&1&1&0&0&0&0\\ 0&0&0&0&1&0&0&0\\ 1&0&1&0&0&1&1&0\\ 0&0&0&1&0&0&0&1\\ 1&0&0&0&0&1&0&0\\ 0&0&1&0&0&0&1&0\\ 1&1&0&0&0&1&0&0\\ 0&0&0&0&1&0&0&0\\ 0&0&0&0&0&0&0&0\\ 0&0&0&1&1&0&0&1\\ 0&0&1&1&1&0&1&1\\ 0&0&0&0&0&0&0&0\\ 0&0&0&0&0&1&1&1\\ 0&1&0&0&0&0&0&0\\ 0&0&0&0&0&0&0&0 \end{bmatrix} \delta^{1,3} = \begin{bmatrix} 1&0&1&1&1&0\\ 0&1&0&0&0&1\\ 0&0&0&0&0&0\\ 0&0&0&0&0&0\\ 1&0&1&0&0&0\\ 0&0&0&1&1&0\\ 0&0&1&0&0&1\\ 1&1&0&0&0&0\\ 0&0&0&0&0&0\\ 0&0&1&0&0&1\\ 0&0&0&0&0&0\\ 0&0&0&1&0&0\\ 0&0&0&0&1&0\\ 1&1&0&1&0&0\\ 0&0&0&0&0&0\\ 1&1&1&1&0&1\\ 0&1&0&0&0&1\\ 0&0&0&0&0&0\\ 0&0&0&0&1&0 \end{bmatrix}$$

$$\delta^{4,2} = \begin{bmatrix} 0&0&0&0&0\\ 0&0&0&0&0\\ 1&0&1&1&1\\ 0&1&0&0&0\\ 1&0&1&1&0\\ 0&0&0&0&1\\ 1&0&1&0&0\\ 0&0&0&1&0\\ 1&0&0&0&0\\ 0&0&1&0&0\\ 1&0&0&0&0\\ 0&1&0&0&1\\ 0&0&0&0&0\\ 0&1&0&1&1\\ 0&1&1&1&1\\ 0&0&0&0&0\\ 0&0&0&0&0\\ 0&0&0&0&0\\ 0&0&0&0&0 \end{bmatrix} \delta^{4,3} = \begin{bmatrix} 0&0&0&0&0&0\\ 0&0&0&0&0&0\\ 1&0&1&1&1&0\\ 0&1&0&0&0&1\\ 1&0&1&0&0&0\\ 0&0&0&1&1&0\\ 0&0&1&0&0&0\\ 1&0&0&0&0&0\\ 0&0&0&0&0&0\\ 0&0&1&0&0&0\\ 0&0&0&0&0&0\\ 0&0&0&1&0&0\\ 0&1&0&0&1&0\\ 1&0&0&1&0&0\\ 0&0&0&0&0&0\\ 1&0&1&1&0&1\\ 0&0&0&0&0&0\\ 0&0&0&0&0&0\\ 0&1&0&0&1&0 \end{bmatrix}$$

As regards electric paths there is 1 origin node (4) and 2 destination nodes (2 and 3). Therefore, there are only two classes of paths, namely P^2_{42} and P^2_{43}. Tables 7.7 and 7.8 enumerates the paths (link sequences) for each class.

For each class of paths, the link to path matrices are:

Table 7.7 Paths for P^2_{42}

#Path	P^2_{42} links				
1	20	21	22	23	11
2	20	21	22	9	11
3	20	21	7	9	11
4	20	5	7	9	11
5	3	21	7	9	11
6	3	5	22	9	11
7	3	5	7	23	11
8	20	21	7	23	11
9	20	5	7	23	11
10	3	21	22	23	11
11	3	5	22	23	11
12	20	5	22	23	11
13	20	21	7	9	11
14	3	21	22	9	11
15	3	21	7	23	11
16	20	21	22	10	15
17	20	21	7	10	15
18	20	5	22	10	15
19	3	21	22	10	15
20	20	5	7	10	15
21	3	21	7	10	15
22	3	5	22	10	15
23	20	21	8	14	15
24	20	5	8	14	15
25	3	21	8	14	15
26	20	6	12	14	15

Table 7.8 Paths for P_{43}^2

#Path	P_{43}^2 links				
1	20	21	8	14	16
2	20	5	8	14	16
3	3	21	8	14	16
4	20	21	22	10	16
5	20	21	7	10	16
6	3	21	22	10	16
7	20	5	22	10	16
8	20	5	7	10	16
9	3	21	7	10	16
10	3	5	22	10	16
11	20	6	12	14	16
12	20	6	13	19	0

$$\xi^{4,3} = \begin{bmatrix} 0\,0\,0\,0 & 0\,0\,0\,0 & 0\,0\,0\,0 \\ 0\,0\,0\,0 & 0\,0\,0\,0 & 0\,0\,0\,0 \\ 0\,0\,1\,0 & 0\,1\,0\,0 & 1\,1\,0\,0 \\ 0\,0\,0\,0 & 0\,0\,0\,0 & 0\,0\,0\,0 \\ 0\,1\,0\,0 & 0\,0\,1\,1 & 0\,1\,0\,0 \\ 0\,0\,0\,0 & 0\,0\,0\,0 & 0\,0\,1\,1 \\ 0\,0\,0\,0 & 1\,0\,0\,1 & 1\,0\,0\,0 \\ 1\,1\,1\,0 & 0\,0\,0\,0 & 0\,0\,0\,0 \\ 0\,0\,0\,0 & 0\,0\,0\,0 & 0\,0\,0\,0 \\ 0\,0\,0\,1 & 1\,1\,1\,1 & 1\,1\,0\,0 \\ 0\,0\,0\,0 & 0\,0\,0\,0 & 0\,0\,0\,0 \\ 0\,0\,0\,0 & 0\,0\,0\,0 & 0\,0\,1\,0 \\ 0\,0\,0\,0 & 0\,0\,0\,0 & 0\,0\,0\,1 \\ 1\,1\,1\,0 & 0\,0\,0\,0 & 0\,0\,1\,0 \\ 0\,0\,0\,0 & 0\,0\,0\,0 & 0\,0\,0\,0 \\ 1\,1\,1\,1 & 1\,1\,1\,1 & 1\,1\,1\,0 \\ 0\,0\,0\,0 & 0\,0\,0\,0 & 0\,0\,0\,0 \\ 0\,0\,0\,0 & 0\,0\,0\,0 & 0\,0\,0\,0 \\ 0\,0\,0\,0 & 0\,0\,0\,0 & 0\,0\,0\,1 \\ 1\,1\,0\,1 & 1\,0\,1\,1 & 0\,0\,1\,1 \\ 1\,0\,1\,1 & 1\,1\,0\,0 & 1\,0\,0\,0 \\ 0\,0\,0\,1 & 0\,1\,1\,0 & 0\,1\,0\,0 \\ 0\,0\,0\,0 & 0\,0\,0\,0 & 0\,0\,0\,0 \end{bmatrix}$$

As regards ω, it is assumed to be equal to 0.1 [h/V], $q^1_{1,2} = 100$, $q^1_{1,3} = 100$, $q^1_{4,2} = 100$, $q^1_{4,3} = 100$, $q^2_{4,2} = 10$, and $q^2_{4,3} = 10$ [#vehicles/h].

7.6.1 Multinomial Logit Results

Optimal results are obtained by using LINGO as a solver of the nonlinear system (7.4)–(7.20) and are reported in Tables 7.9, 7.10, 7.11, 7.12, 7.13 and 7.14. These tables show the total cost, the flow, and the probability for each path, while Table 7.15 shows the flow for each arc.

Table 7.9 Multinomial Logit: Results for P^1_{12}

#Path	$c_{\text{path},k}$	$f_k^{1,2}$	$Q_k^{1,2}$
1	0.9405	12.1257	0.1213
2	0.4705	19.4006	0.1940
3	1.2016	9.3390	0.0934
4	1.2123	9.2395	0.0924
5	1.0540	10.8249	0.1082
6	0.6995	15.4296	0.1543
7	0.9607	11.8836	0.1188
8	0.9714	11.7570	0.1176

Table 7.10 Multinomial Logit: Results for P^1_{13}

#Path	$c_{\text{path},k}$	$f_k^{1,3}$	$Q_k^{1,3}$
1	1.1650	13.6375	0.1364
2	0.9241	17.3533	0.1735
3	1.1543	13.7843	0.1378
4	1.0067	15.9775	0.1598
5	0.7002	21.7074	0.2171
6	0.9134	17.5401	0.1754

Table 7.11 Multinomial Logit: Results for P^1_{42}

#Path	$c_{\text{path},k}$	$f_k^{4,2}$	$Q_k^{4,2}$
1	0.9211	22.1723	0.2217
2	0.8393	24.0624	0.2406
3	1.1823	17.0767	0.1708
4	1.1930	16.8948	0.1689
5	1.0346	19.7938	0.1979

Table 7.12 Multinomial Logit: Results for P^1_{43}

#Path	$c_{\text{path},k}$	$f_k^{4,3}$	$Q_k^{4,3}$
1	1.1457	11.4990	0.1150
2	0.4856	22.2508	0.2225
3	1.1350	11.6228	0.1162
4	0.9873	13.4721	0.1347
5	0.6808	18.3035	0.1830
6	0.4589	22.8519	0.2285

Table 7.13 Multinomial Logit: Results for P^2_{42}

#Path	$\hat{c}_{\text{path},k}$	$g_k^{4,2}$	$\hat{Q}_k^{4,2}$
1	8.3903	0.0037	0.0004
2	6.6744	0.0207	0.0021
3	4.8420	0.1292	0.0129
4	3.1604	0.6943	0.0694
5	2.6028	1.2125	0.1213
6	2.7536	1.0428	0.1043
7	2.6370	1.1717	0.1172
8	6.5579	0.0232	0.0023
9	4.8762	0.1248	0.0125
10	6.1511	0.0349	0.0035
11	4.4694	0.1875	0.0188
12	6.7086	0.0200	0.0020
13	4.8420	0.1292	0.0129
14	4.4352	0.1940	0.0194
15	4.3186	0.2180	0.0218
16	6.9355	0.0159	0.0016
17	5.1031	0.0995	0.0099
18	5.2539	0.0856	0.0086
19	4.6963	0.1494	0.0149
20	3.4215	0.5347	0.0535
21	2.8639	0.9339	0.0934
22	3.0147	0.8032	0.0803
23	5.1138	0.0984	0.0098
24	3.4322	0.5290	0.0529
25	2.8746	0.9239	0.0924
26	3.2738	0.6198	0.0620

Table 7.14 Multinomial Logit: Results for P_{43}^2

#Path	$\hat{c}_{\text{path},k}$	$g_k^{4,3}$	$\hat{Q}_k^{4,3}$
1	5.0665	0.1747	0.0175
2	3.3849	0.9388	0.0939
3	2.8273	1.6395	0.1639
4	6.8882	0.0283	0.0028
5	5.0558	0.1766	0.0177
6	4.6490	0.2652	0.0265
7	5.2066	0.1518	0.0152
8	3.3742	0.9489	0.0949
9	2.8166	1.6571	0.1657
10	2.9674	1.4252	0.1425
11	3.2265	1.0998	0.1100
12	2.9201	1.4943	0.1494

Table 7.15 Multinomial Logit: Optimal flow on arcs

x_a	x_r
106.6357	8.1411
93.3643	8.1278
142.6938	4.4282
69.1651	1.7839
146.0499	
103.2930	
139.0277	
84.6854	
53.1502	
88.5217	
74.3348	
85.8503	
63.7559	
170.5358	
135.6652	
146.2441	
73.9636	
19.4006	
63.7559	

As one can see, the SUE conditions are respected, as a matter of fact, the probability of a path is higher when the cost is lower.

7.6.2 Multinomial C-Logit Results

In order to obtain optimal results in the case of the C-logit probability model, it is necessary to introduce the commonality factor $cf^{sd}_{\text{path},k}$ for each path of the network. These values are given in Tables 7.16, 7.17, 7.18, 7.19, 7.20 and 7.21.

Table 7.16 Commonality factor for P^1_{12}

#Path	$cf^{12}_{\text{path},k}$
1	0.8146
2	0.2554
3	0.9555
4	0.8755
5	0.5878
6	0.84
7	0.9776
8	0.8994

Table 7.17 Commonality factor for P^1_{13}

#Path	$cf^{13}_{\text{path},k}$
1	0.7991
2	0.5878
3	0.7049
4	0.7165
5	0.1116
6	0.47

Table 7.18 Commonality factor for P^1_{42}

#Path	$cf^{42}_{\text{path},k}$
1	0.1823
2	0.2939
3	0.6008
4	0.7165
5	0.6264

Table 7.19 Commonality factor for P_{43}^{1}

#Path	$cf_{\text{path},k}^{43}$
1	0.4991
2	0.1438
3	0.3696
4	0.7505
5	0.3865
6	0.1681

Table 7.20 Commonality factor for P_{42}^{2}

#Path	$c\hat{f}_{\text{path},k}^{42}$
1	2.4681
2	2.451
3	2.451
4	2.3609
5	2.3795
6	2.2824
7	2.3026
8	2.4681
9	2.3795
10	2.3979
11	2.3026
12	2.3795
13	2.451
14	2.3795
15	2.3979
16	2.3795
17	2.3795
18	2.2824
19	2.3026
20	2.2824
21	2.3026
22	2.1972
23	2.1518
24	2.0281
25	2.0541
26	1.6864

Table 7.21 Commonality factor for P_{43}^2

#Path	$\hat{cf}_{\text{path},k}^{43}$
1	1.6908
2	1.6141
3	1.5261
4	1.8283
5	1.7957
6	1.6864
7	1.7619
8	1.727
9	1.6487
10	1.6094
11	1.4463
12	0.5816

As in the previous case, optimal results are obtained by solving the nonlinear system (7.4)–(7.20) and are reported in Tables 7.22, 7.23, 7.24, 7.25, 7.26 and 7.27. These tables show the total cost, the flow, and the probability for each path, while Table 7.28 shows the flow for each arc.

From a careful examination of the obtained results, the introduction of the commonality factor changes notably the distribution of the flows to the equilibrium on the net. Therefore, the path overlapping factor results to have an important influence on the users' decisions.

Table 7.22 Multinomial C-Logit: Results for P_{12}^1

#Path	$c_{\text{path},k}$	$f_k^{1,2}$	$Q_k^{1,2}$
1	1.0169	12.0402	0.1204
2	0.4961	20.2689	0.2027
3	1.2883	9.1783	0.0918
4	1.2910	9.1538	0.0915
5	1.1086	10.9850	0.1098
6	0.7832	15.2100	0.1521
7	1.0543	11.5984	0.1160
8	1.0571	11.5654	0.1157

Table 7.23 Multinomial C-Logit: Results for P_{13}^1

#Path	$c_{\text{path},k}$	$f_k^{1,3}$	$Q_k^{1,3}$
1	1.2364	13.2625	0.1326
2	0.9790	17.1554	0.1716
3	1.2163	13.5316	0.1353
4	1.0745	15.5925	0.1559
5	0.6896	22.9130	0.2291
6	0.9566	17.5449	0.1754

Table 7.24 Multinomial C-Logit: Results for P_{42}^1

#Path	$c_{\text{path},k}$	$f_k^{4,2}$	$Q_k^{4,2}$
1	0.9343	22.7698	0.2277
2	0.8647	24.4107	0.2441
3	1.2334	16.8822	0.1688
4	1.2557	16.5110	0.1651
5	1.0931	19.4263	0.1943

Table 7.25 Multinomial C-Logit: Results for P_{43}^1

#Path	$c_{\text{path},k}$	$f_k^{4,3}$	$Q_k^{4,3}$
1	1.1457	11.3080	0.1131
2	0.4856	23.1187	0.2312
3	1.1350	11.5783	0.1158
4	0.9873	12.8580	0.1286
5	0.6808	18.0378	0.1804
6	0.4589	23.0992	0.2310

Table 7.26 Multinomial C-Logit: Results for P_{42}^2

#Path	$\hat{c}_{\text{path},k}$	$g_k^{4,2}$	$\hat{Q}_k^{4,2}$
1	1.1870	0.0036	0.0004
2	0.4719	0.0201	0.0020
3	1.1634	0.1257	0.0126
4	1.0585	0.6842	0.0684
5	0.7200	1.1920	0.1192
6	0.4727	1.0388	0.1039
7	1.1870	1.1651	0.1165
8	6.5579	0.0226	0.0023
9	4.8762	0.1228	0.0123
10	6.1511	0.0342	0.0034
11	4.4694	0.1865	0.0186
12	6.7086	0.0197	0.0020
13	4.8420	0.1257	0.0126
14	4.4352	0.1908	0.0191
15	4.3186	0.2140	0.0214
16	6.9355	0.0157	0.0016
17	5.1031	0.0979	0.0098
18	5.2539	0.0853	0.0085
19	4.6963	0.1486	0.0149
20	3.4215	0.5331	0.0533
21	2.8639	0.9287	0.0929
22	3.0147	0.8100	0.0810
23	5.1138	0.0991	0.0099
24	3.4322	0.5411	0.0541
25	2.8746	0.9419	0.0942
26	3.2738	0.6528	0.0653

Table 7.27 Multinomial C-Logit: Results for P_{43}^2

#Path	$\hat{c}_{\text{path},k}$	$g_k^{4,3}$	$\hat{Q}_k^{4,3}$
1	5.2334	0.1702	0.0170
2	3.5406	0.9248	0.0925
3	2.9748	1.6284	0.1628
4	7.0689	0.0271	0.0027
5	5.2332	0.1702	0.0170
6	4.8126	0.2592	0.0259
7	5.3771	0.1474	0.0147
8	3.5412	0.9242	0.0924
9	2.9764	1.6258	0.1626
10	3.1197	1.4087	0.1409
11	3.3702	1.0966	0.1097
12	2.9817	1.6173	0.1617

Table 7.28 Multinomial C-Logit: Optimal flow on arcs

x_a	x_r
106.6569	8.2273
93.3431	8.0416
141.1441	4.3957
70.6286	1.7684
144.8075	
103.1792	
138.2658	
83.2617	
53.3974	
87.4957	
75.4347	
85.0218	
65.6868	
168.2835	
134.5653	
144.3132	
73.0742	
20.2689	
65.6868	

References

1. Aicardi M, Ferro G, Minciardi R, Robba M (2023) Stochastic traffic assignment with electric vehicles: a convex optimization approach. IFAC-PapersOnLine 56/2:9074–9079
2. Ferro G, Minciardi R, Robba M (2020) A user equilibrium model for EV: joint traffic and energy demand assignment. Energy 198:117299
3. Daganzo CF, Sheffi Y (1977) On stochastic models of traffic assignment. Transp Sci 11(3):253–274
4. Bliemer MCJ, Bovy PHL (2008) Impact of route choice set on route choice probabilities. Transp Res Rec 2076(1):10–19
5. Bekhor S, Prato CG (2006) Effects of choice set composition in route choice modeling. In: 11th conference of the international association for travel behaviour research
6. Zhou Z, Sun H, Guo Q (2018) Stochastic user equilibrium in charging station selection based on discrete choice model. In: 2018 IEEE power & energy society general meeting (PESGM). IEEE
7. Riemann R, Wang DZW, Busch F (2015) Optimal location of wireless charging facilities for EV: flow-capturing location model with stochastic user equilibrium. Transp Res Part C Emerg Technol 58:1–12
8. He F et al (2016) Sustainability SI: optimal prices of electricity at public charging stations for plug-in EV. Netw Spatial Econ 16(1):131–154
9. He F, Yin Y, Zhou J (2013) Integrated pricing of roads and electricity enabled by wireless power transfer. Transp Res Part C Emerg Technol 34:1–15

10. Ferro G, Minciardi R, Parodi L, Robba M (2020) A bi-level approach for the optimal planning of charging stations and electric vehicles traffic assignment. In: 2020 7th international conference on control, decision and information technologies (CoDIT), Prague, Czech Republic, 533–538
11. United States. Bureau of Public Roads (1964) Traffic assignment manual for application with a large, high speed computer, vol 37. US Department of Commerce, Bureau of Public Roads, Office of Planning, Urban Planning Division
12. Davidson KB (1966) A flow travel time relationship for use in transportation planning. In: Australian road research board (ARRB) conference, 3rd, Sydney, vol 3, no 1
13. Sheffi Y (1985) Urban transportation networks, vol 6. Prentice-Hall, Englewood Cliffs, NJ
14. Cascetta E et al (1996) A modified logit route choice model overcoming path overlapping problems. Specification and some calibration results for interurban networks. In: Transportation and traffic theory. Proceedings of The 13th international symposium on transportation and traffic theory, Lyon, France, 24–26 July 1996
15. Nguyen S, Dupuis C (1984) An efficient method for computing traffic equilibria in networks with asymmetric transportation costs. Transp Sci 18(2):185–202

Chapter 8
Planning for Electric Vehicles: A Stochastic User Equilibrium Approach for Joint Traffic Assignment and Sizing of Charging Stations

8.1 Introduction[1]

In this chapter, attention is focused on the optimal sizing and siting of electric vehicle charging stations (EVCSs). The planning of EVCSs is formalized as an optimization problem where stochastic user equilibrium (SUE), derived in Chap. 7, are constraints of the whole problem. In the problem formalization, we will assume that the Logit-based SUE can describe all drivers' behavior (including both the drivers of conventional vehicles and EVs). In previous works [1–3], we studied a similar decision problem, but the behavior of all drivers was assumed to follow the deterministic UE conditions for electric vehicles (EVs) [4]. This issue is very important from a practical point of view since a more realistic model for the drivers' behavior is adopted.

Current research literature studies the planning of EVCSs in a multidisciplinary way since it involves different aspects such as the estimation of charging demand, the transportation demand, drivers' behavior, the presence of specific areas that are more suitable for EVCSs location (such as commercial centers, industrial areas), etc. Our work is not so general to consider all these issues, but it is comparable with many works assembling different expertise to solve the problem of EVCSs planning.

An important set of contributions in the literature considers the simultaneous optimal planning of EVCSs and DG (Distributed Generation) power units in distribution systems [4–7]. These issues will be considered in Chap. 9.

Another collection of papers is related to the study of the relationship between the individual drivers' choices and the EVs' charging demand. This objective is achieved by extending the classical SUE conditions [8] to mixed fleets (both traditional vehicles and EVs) [9–12]. This is because SUE traffic assignment influences the energy demand distribution and thus the charging stations' planning. For this reason, bi-level decision architectures are very popular in current literature, as in [1, 13, 14]. In [15], the authors focus their attention on computing paths thanks to a dynamic traffic flow

[1] Text in this section is © 2020 IEEE. Reproduced from [3], https://doi.org/10.1109/CoDIT49905.2020.9263928; used with permission.

G. Ferro et al., *Optimization of Electric-Vehicle Charging*, Advances in Industrial Control, https://doi.org/10.1007/978-3-031-61917-5_8

simulation, allowing to consider real-time traffic conditions and at the same time minimizing the costs for the optimal planning of EVCS. In [16], two different kinds of networks have been considered: the traffic network (TN) and the power distribution network (PDN), performing a complete planning of both network elements including also EVCS. Zhou et al. [17] compare the different planning results with respect to three UE problems to study the influence of uncertainty using Stochastic System Optimum, System Optimum, and Traffic Optimum. In [18], a two-stage planning problem for EVCSs placement is addressed. In the first stage, candidate locations for EVCSs are identified thanks to the use of fuzzy logic strategy, considering different features related to TN and PDN. Therefore, the optimization is performed to choose the optimal locations, and design of EVCSs. Hu et al. [19] propose dynamical traffic assignment model with time-varying demand. UE conditions to estimate the demand in different periods for the optimal power flow are therefore taken into account. Finally, [20] proposes a holistic approach for the design and location of fast EVCSs that considers TN, PDN, installation costs, territorial characteristics, and daily traffic conditions.

This chapter defines a bi-level optimization problem that uses the traffic model to derive the SUE based on a stochastic user's choice model obtained from the Logit model. Moreover, the size, the site, and the energy price of the EVCSs will be the object of optimization. The possible paths are calculated based on present charging stations and new potential charging facilities.

8.2 Formalization of the Optimization Problem[2]

Sets

- A: directed links;
- S: traffic sources;
- D: traffic destination;
- P^1_{sd}: set of paths for traditional vehicles and EVs that do not need to recharge;
- K^1_{sd}: number of paths of set P^1_{sd}, P^1_{sd}, $k = 1, \ldots, K^1_{sd}$;
- P^2_{sd}: set of paths for EVs that need a recharge during a trip for currently installed EVCSs;
- K^2_{sd}: number of paths of set P^2_{sd}, P^2_{sd}, $k = 1, \ldots, K^2_{sd}$;
- $\tilde{P}^2_{sd}$: paths for EVs that need a recharge during a trip for new *potential* EVCSs;
- $\tilde{K}^2_{sd}$: number of paths of set $\tilde{P}^2_{sd}$, $k = 1, \ldots, \tilde{K}^2_{sd}$;
- $\tilde{R}_1$: EVCSs belonging to the previously established competitors;
- $\tilde{R}_2$: *potential* EVCS belonging to the new competitor;
- $\tilde{R} = \tilde{R}_1 \cup \tilde{R}_2$.

[2] Text in this section is © 2020 IEEE. Reproduced from [3], https://doi.org/10.1109/CoDIT49905.2020.9263928; used with permission.

The set of optimization variables of the problem are:

- y_i (binary variable) equal to 1 if the i-th potential charging station is selected and 0 otherwise;
- CAPs_i i.e., the service capacity, of the new CS $i \in \tilde{R}_2$;
- gain_i [€], i.e., the unit service price for the customers of the EVCS stations of the new competitor.

$$\text{pr}_i = \text{const} + \text{gain}_i \quad i \in \tilde{R}_2 \tag{8.1}$$

where const is a (fixed) additive constant which represents the fixed cost of the energy provided to a single customer.

The decision variables set represent the behavior of the traffic system, including x_i, as the flow over link $i \in \tilde{R}$ measured in [#veh/h] or in [#services/h].

The objective function of the new competitor for the installation of new EVCS in $i \in \tilde{R}_2$ is the minimization of its net yearly cost:

$$\min \sum_{i \in \tilde{R}_2} \left(k_i y_i + h_i \text{CAPs}_i - N_{\text{Hours}} \text{gain}_i y_i x_i \right) \tag{8.2}$$

where:

- k_i is the installation fixed cost for a new EVCS [€/year];
- h_i is the proportionality coefficient of the variable cost for the installation of a new EVCS [€/year(#veh/h)];
- N_{Hours} is the total hours in a year.

The following operative constraints are considered. First, it is supposed that the maximum number of new EVCS is less than equal to H:

$$\sum_{i \in \tilde{R}_2} y_i \leq H. \tag{8.3}$$

Furthermore, there are bound constraints and disjunctive constraints that correlate binary and continuous variables. That is:

$$My_i - \text{CAPs}_i \geq 0 \quad i \in \tilde{R}_2 \tag{8.4}$$

$$0 \leq \text{CAPs}_i \leq \text{CAPs}_i^{\text{MAX}} \quad i \in \tilde{R}_2 \tag{8.5}$$

$$0 \leq x_i \leq \text{CAPs}_i \quad i \in \tilde{R}_1 \tag{8.6}$$

$$My_i - x_i \geq 0 \quad i \in \tilde{R}_2 \tag{8.7}$$

where constraints (8.4) and (8.8) are the disjunctive relationships that impose the payment of the fixed cost in the presence of a new EVCS (with nonzero service capacity), constraints (8.5) are a priori conditions over the size of the new EVCS, constraints (8.6) and (8.7) define the maximum EVCS capacity.

A second set of constraints is relevant to the traffic flow model. To this end, it is necessary to distinguish between the demand of traditional vehicles and the demand of EVs.

We define two kinds of traffic requests supposed to be known and to be satisfied for any origin–destination pair (s, d): q^1_{sd}, i.e., traditional vehicles or EVs which do not need a recharge during the trip; q^2_{sd}, i.e., EVs that need to recharge during the trip:

$$q^1_{sd} = \sum_{k=1}^{K^1_{sd}} f_k^{sd} \quad \forall (s, d) \in O/D \tag{8.8}$$

$$q^2_{sd} = \sum_{k=1}^{K^2_{sd}} g_k^{sd} + \sum_{k=1}^{\tilde{K}^2_{sd}} \tilde{g}_k^{sd} \quad \forall (s, d) \in O/D \tag{8.9}$$

where:

- f_k^{sd} is the flow relevant to vehicles that do not need to recharge during the trip in P^1_{sd}, $k = 1, \ldots, K^1_{sd}$;
- g_k^{sd} and $\tilde{g}_k^{sd}$ are the flows relevant to EVs that need to recharge during the trip [veh/h] in P^2_{sd}, $k = 1, \ldots, K^2_{sd}$ and $\tilde{P}^2_{sd}$, $k = 1, \ldots, \tilde{K}^2_{sd}$, respectively.

To satisfy traffic flow constraints it is necessary to relate traffic flows over a link to each path. The overall traffic flow x_a on a link $a \in A$ is the sum of the flows f_k^{sd} corresponding to paths in P^1_{sd} that include the link $a \in A$, plus the flow $g_k^{sd}(\tilde{g}_k^{sd})$ corresponding to paths in $P^2_{sd}(\tilde{P}^2_{sd})$ that include link $a \in A$. Instead, by definition, for a link $a \in \tilde{R}$ there is no flow f_k^{sd} over this link. That is:

$$x_a = \sum_{(s,d)\in O/D} \sum_{k=1}^{K^1_{sd}} f_k^{sd} \delta_{a,k}^{sd} + \sum_{(s,d)\in O/D} \sum_{k=1}^{K^2_{sd}} g_k^{sd} \xi_{a,k}^{sd} + \sum_{(s,d)\in O/D} \sum_{k=1}^{\tilde{K}^2_{sd}} \tilde{g}_k^{sd} \tilde{\xi}_{a,k}^{sd} \quad a \in A \tag{8.10}$$

with

$$\delta_{a,k}^{sd} = \begin{cases} 1 & \text{if link } a \in \text{the } k\text{-th path in } P^1_{sd} \\ 0 & \text{otherwise} \end{cases} \quad a \in A, (s, d) \in O/D, k = 1, \ldots, K^1_{sd} \tag{8.11}$$

$$\xi_{a,k}^{sd} = \begin{cases} 1 & \text{if link } a \in \text{to the } k\text{-th path in } P^2_{sd} \\ 0 & \text{otherwise} \end{cases} \quad a \in A \cup \tilde{R}_1, (s, d) \in O/D, k = 1, \ldots, K^2_{sd} \tag{8.12}$$

$$\tilde{\xi}_{a,k}^{sd} = \begin{cases} 1 & \text{if link } a \in \text{to the } k\text{-th path in } \tilde{P}_{sd}^2 \\ 0 & \text{otherwise} \end{cases} \quad a \in A \cup \tilde{R}_2, (s,d) \in O/D, k = 1, \ldots, \tilde{K}_{sd}^2 \tag{8.13}$$

$$x_a = \sum_{(s,d)\in O/D} \sum_{k=1}^{K_{sd}^2} g_k^{sd} \xi_{a,k}^{sd} \quad a \in \tilde{R}_1 \tag{8.14}$$

$$x_a = \sum_{(s,d)\in O/D} \sum_{k=1}^{\tilde{K}_{sd}^2} \tilde{g}_k^{sd} \tilde{\xi}_{a,k}^{sd} \quad a \in \tilde{R}_2. \tag{8.15}$$

The definition of costs over each path are

$$c_{\text{path},k}^{sd} = \sum_{a\in A} t_a(x_a)\delta_{a,k}^{sd} \quad k = 1, \ldots, K_{sd}^1 \tag{8.16}$$

$$\hat{c}_{\text{path},k}^{sd} = \sum_{a\in A} t_a(x_a)\xi_{a,k}^{sd} + \sum_{a\in\tilde{R}_1} c_{\text{tot},a}(x_a, y_a)\xi_{a,k}^{sd} \quad k = 1, \ldots, K_{sd}^2 \tag{8.17}$$

$$\tilde{c}_{\text{path},k}^{sd} = \sum_{a\in A} t_a(x_a)\tilde{\xi}_{a,k}^{sd} + \sum_{a\in\tilde{R}_2} c_{\text{tot},a}(x_a, y_a)\tilde{\xi}_{a,k}^{sd} \quad k = 1, \ldots, \tilde{K}_{sd}^2 \tag{8.18}$$

with

$$t_a(x_a) = t_a^0\left[1 + \alpha_a\left(\frac{x_a}{\text{CAP}_a}\right)^4\right] \quad a \in A \tag{8.19}$$

$$c_{\text{tot},a}(x_a, y_a) = t_a(x_a) + ts_a^{sd,0} + \beta_a\left(\frac{y_a}{\text{CAPs}_a - y_a + \zeta_a}\right) + \omega p_{r,a}\overline{\text{ER}} \quad a \in \tilde{R}, \forall (s,d) \in O/D. \tag{8.20}$$

From the previous relationships we define for each link a: t_a^0 as the free-flow time [h]; α_a is a tuning parameter based on field data; CAP_a [#veh/h] is the maximum flow; $ts_a^{sd,0}$ is the EVCS service time [h], β_a is an a priori determined parameter [h] distinctive of the EVCS; CAPs_a is the capacity of the EVCS [# of services/h]; $p_{r,a}$ is the unit rate [€/kWh] at the EVCS; ω is a conversion coefficient that allows dimensional consistency [h/€]; ζ_a is a small number that is introduced to avoid unbounded cost; $\overline{\text{ER}}$ represents the energy request for EVs.

Finally, we introduce constraints representing the SUE conditions described in the previous chapter. To this end, we introduce the road choice probabilities $\hat{Q}_k^{sd}$ and $\tilde{Q}_k^{sd}$, Q_k^{sd}, for the three classes of routes. Here again, the probability that a given route is chosen (within a given class of routes) is the probability that its travel time is perceived to be the lowest of all the routes.

That is:

$$f_k^{sd} = Q_k^{sd} q_{sd}^1 \quad \forall (s, d) \in O/D, k \in K_{sd}^1 \tag{8.21}$$

$$g_k^{sd} = \hat{Q}_k^{sd} q_{sd}^2 \quad \forall (s, d) \in O/D, k \in K_{sd}^2 \tag{8.22}$$

$$\tilde{g}_k^{sd} = \tilde{Q}_k^{sd} q_{sd}^2 \quad \forall (s, d) \in O/D, k \in \tilde{K}_{sd}^2. \tag{8.23}$$

Using as users' choice function the multinomial Logit, it is possible to write the path's probabilities as a function of costs over paths:

$$Q_k^{sd} = \frac{e^{-c_{\text{path},k}^{sd}}}{\sum_{k \in K_{sd}^1} e^{-c_{\text{path},k}^{sd}}} \quad \forall (s, d) \in O/D, k \in K_{sd}^1 \tag{8.24}$$

$$\hat{Q}_k^{sd} = \frac{e^{-\hat{c}_{\text{path},k}^{sd}}}{\sum_{k \in K_{sd}^2} e^{-\hat{c}_{\text{path},k}^{sd}}} \quad \forall (s, d) \in O/D, k \in K_{sd}^2 \tag{8.25}$$

$$\tilde{Q}_k^{sd} = \frac{e^{-\tilde{c}_{\text{path},k}^{sd}}}{\sum_{k \in \tilde{K}_{sd}^2} e^{-\tilde{c}_{\text{path},k}^{sd}}} \quad \forall (s, d) \in O/D, k \in \tilde{K}_{sd}^2. \tag{8.26}$$

Thus, the overall optimization problem carries out the minimization of (8.2), subject to constraints (8.1) and (8.3)–(8.26).

8.3 Application to a Case Study[3,4]

In particular, the Liguria Region has been considered with one origin (Sampierdarena) and one destination (Recco). The possible travels are (see Fig. 8.1) through the highway (Sampierdarena-Genoa East-Nervi-Recco), in which no charging stations are supposed to be present, and through the urban roads (Sampierdarena-Boccadasse-Quinto-Recco). Moreover, when in Quinto, it is possible to take the highway at Nervi. In Fig. 8.1, the charging stations already present are in links $(\tilde{d}, \tilde{e}, \tilde{g})$. Possible charging stations can be placed in arcs $(\tilde{a}, \tilde{b})$, which connect the same nodes of $(\tilde{a}, \tilde{b}) \in \tilde{R}_2$, till a maximum number of $H = 1$. As regards ω, it is assumed to be equal to 0.1 [h/€], while $q_{sd}^1 = 1000$ and $q_{sd}^2 = 80$ [#vehicles/h], and for the objective function, $k_i = [2 \;\; 20]$, $h_i = [1 \;\; 10]$, and const $= 10$. Other parameters useful for the system's model are reported in Table 8.1.

In the considered case, $A = (a, b, c, d, e, g, l)$, $\tilde{R}_1 = (\tilde{d}, \tilde{e}, \tilde{g})$, and $\tilde{R}_2 = (\tilde{a}, \tilde{b})$.

---

[3] Text in Sect. 8.3 is reproduced from [2]; released under a CC BY-NC-ND license; https://doi.org/10.1016/j.ifacol.2020.12.157.
© 2023 2020 G. Ferro, R. Minciardi, L. Parodi, M. Robba. Reproduced with permission from [2].

[4] Text in Sect. 8.3 is © 2020 IEEE. Reproduced from [3], https://doi.org/10.1109/CoDIT49905.2020.9263928; used with permission.

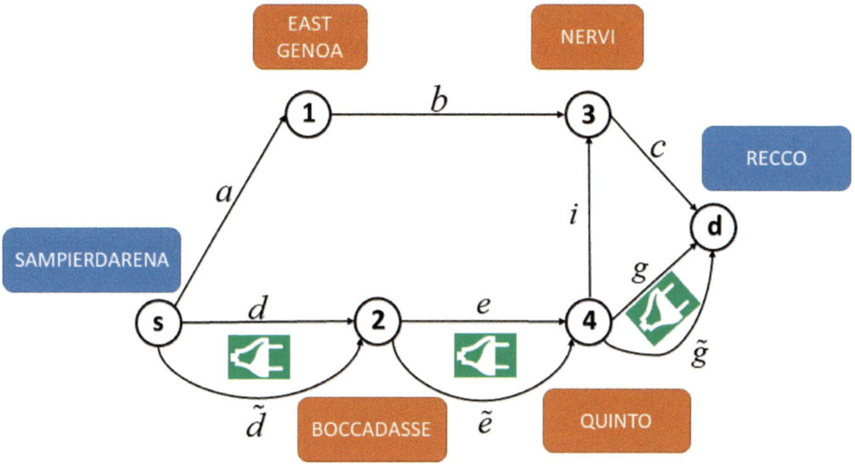

Fig. 8.1 The considered case study; © 2023 IEEE. Reprinted, with permission, from [3]. © 2020 G. Ferro, R. Minciardi, L. Parodi, M. Robba. Reproduced with permission from [2]

Table 8.1 The considered parameters. © 2023 IEEE. Reprinted, with permission, from [3]. See also © 2020 G. Ferro, R. Minciardi, L. Parodi, M. Robba. Reproduced with permission from [2]

	t_a^0 [h]	CAP$_a$ [#vehicles/h]	ts_a^0 [h]	CAPs$_a$ [#vehicles/h]	$p_{r,a}$ [€]
a	0.1	1000	0	0	0
b	0.17	1000	0	0	0
c	0.08	1000	0	0	0
d	0.15	750	0	0	0
e	0.25	400	0	0	0
g	0.25	500	0	0	0
l	0.08	500	0	0	0
$\tilde{d}$	0.2	750	0.5	45	15
$\tilde{e}$	0.25	400	0.3	55	15
$\tilde{g}$	0.25	500	0.3	60	15
$\tilde{a}$	0.3	500	0.4		
$\tilde{b}$	0.3	400	0.3		

It is necessary to define the components of the different sets related to paths. In particular, the set P_{sd}^1 represents all possible paths that include only links belonging to set A. It can be easily seen by Fig. 8.1 that $K_{sd}^1 = 3$ and $P_{sd}^1 = \{(a, b, c), (d, e, g), (d, e, l, c)\}$. The set P_{sd}^2 (with $K_{sd}^2 = 5$) is calculated by enumerating all possible paths that include one of the links $(\tilde{d}, \tilde{e}, \tilde{g})$. That is

$$P_{sd}^2 = \left\{(d, e, \tilde{g}), (\tilde{d}, e, g), (\tilde{d}, e, l, c), (d, \tilde{e}, g), (d, \tilde{e}, l, c)\right\}.$$

Finally, $\tilde{P}^2_{sd}$ (with $\tilde{K}^2_{sd} = 2$) is calculated enumerating all possible paths that include $(\tilde{a}, \tilde{b}) \in \tilde{R}_2$: $\tilde{P}^2_{sd} = \{(a, \tilde{b}, c), (\tilde{a}, b, c)\}$. In particular, Fig. 8.2 reports the new eligible links and related paths for $\tilde{K}^2_{sd}$ and related to $(\tilde{a}, \tilde{b}) \in \tilde{R}_2$. Other parameters are reported in Table 8.1.

A local optimal solution is found after 1 min while the global solution is found after 20 min. The optimal capacity of the EVCSs is 10 in both locations.

The detailed results related to the transportation flows, costs, and probabilities for each path over the network are reported in Table 8.2, while Table 8.3 reports the optimal flows over links.

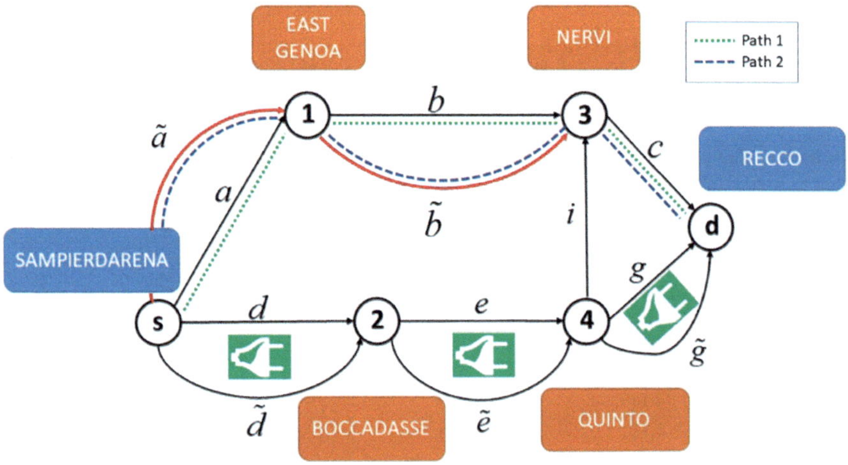

Fig. 8.2 New eligible links (red) and related paths for $\tilde{K}^2_{sd}$ and related to $(\tilde{a}, \tilde{b}) \in \tilde{R}_2$

Table 8.2 Optimal results for the paths in P^2_{sd}

$k \in P^2_{sd}$	g^{sd}_k	$\tilde{c}^{sd}_{\text{path},k}$	Prob. g^{sd}_k	$k \in P^1_{sd}$	f^{sd}_k	$c^{sd}_{\text{path},k}$	Prob. Q^{sd}_k
1	11.3	6.4	0.14	1	695	3.7	0.7
2	6.5	6.9	0.08	2	204	4.9	0.2
3	3.2	7.7	0.04	3	101	5.6	0.1
4	29.2	5.4	0.37				
5	14.5	6.1	0.18	$k \in \tilde{P}^2_{sd}$	$\tilde{g}^{sd}_k$	$\tilde{c}^{sd}_{\text{path},k}$	Prob. $\tilde{Q}^{sd}_k$
				1	8.1	6.7	0.1
				2	7.1	6.9	0.09

Table 8.3 Flows over links in the optimal solution

Link	x_a	Link	x_a
a	703	$\tilde{d}$	9.7
b	702	$\tilde{e}$	43.7
c	829	$\tilde{g}$	11.3
d	360	$\tilde{a}$	7
e	326	$\tilde{b}$	8
g	249		
l	119		

References

1. Ferro G, Minciardi R, Parodi L, Robba M (2021) Optimal planning of charging stations in coupled transportation and power networks based on user equilibrium conditions. IEEE Trans Autom Sci Eng
2. Ferro G, Minciardi R, Parodi L, Robba M (2020) Optimal planning of charging stations and electric vehicles traffic assignment: a bi-level approach. IFAC-PapersOnLine 53(2):13275–13280
3. Ferro G, Minciardi R, Parodi L, Robba M (2020) A bi-level approach for the optimal planning of charging stations and electric vehicles traffic assignment. In: 2020 7th international conference on control, decision and information technologies (CoDIT), pp 533–538
4. Battapothula G, Yammani C, Maheswarapu S (2021) Multi-objective optimal planning of fast charging stations by considering various load models in distribution system. Int J Emerg Electr Power Syst 22(4):439–450
5. Erdinic O, Tascikaraoglu A, Paterakis N, Dursun I, Sinim MC, Catalao JPS (2018) Comprehensive optimization model for sizing and siting of DG units, EV charging stations, and energy storage systems. IEEE Trans Smart Grid 4:3871–3882
6. Simorgh H, Doagou-Moja H, Razmi H, Gharehpetian G (2018) Cost-based optimal siting and sizing of electric vehicle charging stations considering demand response programmes. IET Gener Transm Distrib 8:1712–1720
7. De Angelis E, Carnevale C, Marcoberardino G, Turrini E, Volta M (2021) Low emission road transport scenarios: an integrated assessment of energy demand, air quality, GHG emissions, and costs. IEEE Trans Autom Sci Eng
8. Sheffi Y (1985) Urban transportation networks. Prentice-Hall, Englewood Cliffs, NJ, USA
9. He F, Yin Y, Lawphongpanich S (2014) Network equilibrium models with battery electric vehicles. Transp Res Part B Methodol 67:306–319
10. Cen X, Lo HK, Li L, Lee E (2018) Modeling electric vehicles adoption for urban commute trips. Transp Res Part B Methodol 117(part A):431–454
11. Ferro G, Minciardi R, Robba M (2020) A user equilibrium model for electric vehicles: joint traffic and energy demand assignment. Energy 198:117299
12. Riemann R, Wang DZW, Busch F (2015) Optimal location of wireless charging facilities for electric vehicles: flow-capturing location model with stochastic user equilibrium. Transp Res Part C Emerg Technol 58(part A):1–12
13. Lee YG, Kim HS, Kho SY, Lee C (2014) User-equilibrium-based location model of rapid charging stations for electric vehicles with batteries that have different states of charge. Transp Res Rec: 97–106
14. Xiang Y, Liu J, Li R, Li F, Gu C, Tang S (2016) Economic planning of electric vehicle charging stations considering traffic constraints and load profile templates. Appl Energy 178:647–659
15. Qian Z, Yi Z, Zhong W, Yue H, Yaojia S (2020) Siting and sizing of electric vehicle fast-charging station based on quasi-dynamic traffic flow. IET Renew Power Gener 19:4204–4214

16. Wang X, Shahidehpour M, Jiang C, Li Z (2019) Coordinated planning strategy for electric vehicle charging stations and coupled traffic-electric networks. IEEE Trans Power Syst: 268–279
17. Zhou Z, Sun H, Guo Q (2018) Stochastic user equilibrium in charging station selection based on discrete choice model. In: Presented at 2018 IEEE power & energy society general meeting (PESGM). https://doi.org/10.1109/pesgm.2018.8586008Corpus
18. Deb S, Tammi K, Gao X, Kaȍita K, Mahanta P, Cross S (2021) A robust two-stage planning model for the charging station placement problem considering road traffic uncertainty. IEEE Trans Intell Transp Syst
19. Hu L, Wei W, Shen Z (2019) Impact of time-shiftable traffic demands on coupled transportation and power distribution systems. In: Proceedings of IEEE innovative smart grid technologies—Asia (ISGT Asia), Oct 2019
20. Islam M, Shareef H, Mohamed A (2018) Optimal location and sizing of fast charging stations for electric vehicles by incorporating traffic and power networks. IET Intel Transport Syst 12:947–957

Chapter 9
Planning for Electric Vehicles: Optimal Placement of Charging Stations in an Electrical Distribution Grid

9.1 Introduction[1]

In this chapter, attention is focused on the formalization and implementation of an optimization model for the optimal sizing and siting of renewable generators (e.g., photovoltaic (PV) plants), combined heat and power (CHP) plants, small-scale wind turbines (WT) and electric vehicles (EVs) charging stations over a power distribution grid. It is important to note that this work takes its cue from [1].

The location and sizing of plants for energy production and EVs' charging stations are fundamental when addressing sustainability and emission reduction. When the power distribution grid is considered, it is necessary to guarantee an acceptable quality of service (in terms of power quality and energy request from EVs) while considering the whole lifetime of generation units and charging stations. Generally, this is a complex task, and in literature, different simulation and optimization models are used to support decisions. Such tools are very useful for the decision-making of both Distribution System Operators (DSO) and municipalities that aim to include new facilities within sustainable districts. From this point of view, the approach proposed in this chapter can be seen, rather than a methodology for the design of a single microgrid or plant, as a first attempt of developing a planning tool at a city-level to support a central decision-maker (either public or private) in planning investments for distributed generation, storage systems, charging stations, etc. In the current literature, there are many papers regarding this topic. For instance, [2] conducted a techno-economic optimization using a genetic algorithm approach to determine the optimal size of renewables for an industrial site. [3] Statistical modeling based on the vine-copula theory was utilized to determine EVs' optimal charging station size without considering economic aspects. [4] investigated DGs' optimal location and size to minimize the active distribution network's total cost. The problem was solved using the CVX platform and GUROBI solver. The chaos-embedded SOS algorithm

[1] Text in this section is ©2016 IEEE. Reproduced from [1], https://doi.org/10.1109/ISC2.2016.7580795; used with permission.

G. Ferro et al., *Optimization of Electric-Vehicle Charging*, Advances in Industrial Control, https://doi.org/10.1007/978-3-031-61917-5_9

was employed to determine DGs' optimal location and sizing to decrease power loss and enhance voltage stability in a radial distribution network [5]. The literature includes various objectives, such as cost minimization, for the optimization problem. To reduce the total cost, the optimal sizes of PV, WT, and ESS are specified within [6–8]. Another paper [9] presents an optimization model that uses a second-order conic programming problem to size and locate renewable energy resources (RERs), electric vehicle charging stations, and energy storage systems (ESSs). In [10] a novel method is addressed for sizing batteries in order to reduce load uncertainty while considering the intermittent nature of wind. In all previous papers, the planning of charging stations is not carried out, whereas it is considered in [11, 12]. In [11], the authors propose an algorithm to plan the location of EVs' charging stations over a distribution grid to increase the hosting capacity and offer services in the balancing market, while in [12] the calculation for several fast charging stations in a residential area and an optimization model of fast charging station planning are proposed. Ant colony optimization is implemented to minimize the total costs of fast charging stations and transmission line losses, subject to traffic and power system security constraints.

In this chapter, a decision model is formalized. The aim is to find the optimal location and size of several power production plants and EVs' charging stations to satisfy electrical loads and charging demands in a power distribution network. The considered technologies are small-size WTs, PV, and cogeneration micro-turbines connected to the same grid, and the overall net present value is minimized. The optimal results are obtained over a modified IEEE 13 bus network.

The following section defines the problem, explaining all the technologies considered for the planning problem. Then, Sect. 9.3 describes the modeling of constraints and cost for each device considered for the planning problem. Finally, in Sect. 9.4, the formulation of the planning problem is tested on a modified IEEE 13 bus network.

9.2 Problem Definition[2]

The goal of the planning problem is to find the best sizing, location, and combination of the available technologies over a specific geographical area interconnected by a power distribution grid. The objective function to be minimized is characterized by the sum of the overall installation and operational costs over the considered time horizon of Y years. Hourly and seasonal variations of photovoltaic production, electrical demand, energy prices, etc., have been considered monthly, and, for each month, a typical average day is determined from real data. Each plant and electrical demand are considered as connected to one node of the electrical grid, except for the EVs' charging demand, which is specified for a geographic area (zones) that may contain more than one distribution network node (as reported in Fig. 9.1).

[2] Text in this section is ©2016 IEEE. Reproduced from [1], https://doi.org/10.1109/ISC2.2016.7580795; used with permission.

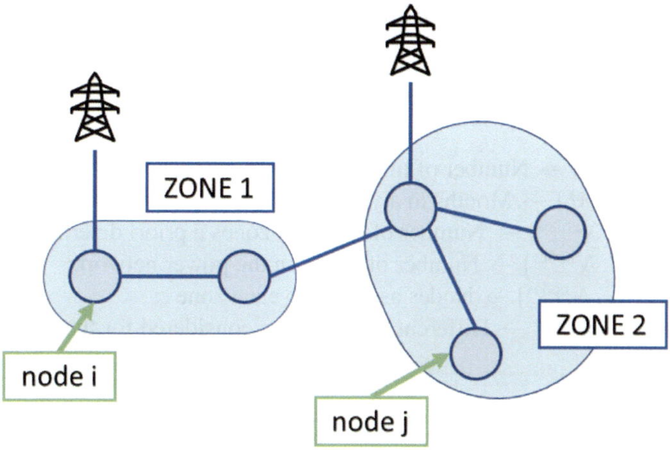

Fig. 9.1 Example of distribution network with different zones

It is important to note that we use this specific modeling since the nodes of the electric grid do not correspond to those of the transportation network that defines the corresponding EVs' charging demand.

We consider the following technologies for the planning problem:

- Photovoltaic plants, whose installation size corresponds to the total panels' surface. The available data on average radiation allows deriving a site-specific generated power potential per area unit, which is multiplied by the decision variable (i.e., photovoltaic surface).
- CHP units, capable of producing both electrical and thermal power. This study considers different CHP models, and the hourly scheduling of each CHP is considered as assigned on each average day, according to the node thermal request. The decision variable related to this technology is the number of installed CHPs in each distribution grid node.
- WTs. The generation potential of a single turbine is estimated based on the wind speed statistics, and the number of installed turbines at each distribution grid node represents the decision variable of the problem.
- EVs' charging stations are needed to fulfill the EVs' charging demand. The decision variable is related to the CS's size and location for each distribution grid node.

9.3 Mathematical Formulation

In order to perform the optimal planning over a distribution network, the following sets have been defined:

- $\tilde{T} = \{1, \ldots, T\} \rightarrow$ Number of time intervals in a day;
- $\tilde{M} = \{1, \ldots, M\} \rightarrow$ Months in a year;
- $\tilde{Z} = \{1, \ldots, N^{\text{zones}}\} \rightarrow$ Number of different zones a priori determined;
- $N = \{1, \ldots, N^{\text{nodes}}\} \rightarrow$ Number of nodes in the power network;
- $N_z = \{1, \ldots, N_z^{\text{nodes}}\} \rightarrow$ Nodes assigned to each zone z;
- $\tilde{\Psi} = \{1, \ldots, N^{\text{tech}}\} \rightarrow$ Different technologies considered for the CS.

9.3.1 Charging Stations Modeling

First of all, we introduce the modeling of the charging stations. These devices are not considered to be assigned to a specific node of the network, but to a geographical area that includes different nodes. The following equation is used to define the size and location of the CS total active power $P_{i,m,t}^{CS}$ in node i of zone z of month m:

$$P_{i,m,t}^{CS} = \sum_{\psi \in \tilde{\Psi}} P_{\psi,i,m,t} n_{\psi,i} \quad i \in N_z, z \in \tilde{Z}, m \in \tilde{M}, t \in \tilde{T} \tag{9.1}$$

where $P_{\psi,i,m,t}$ is the known rated power for each CS technology ψ in node i of zone z of month m, and $n_{\psi,i}$ is an integer decision variable that defines the number of installed CS with technology ψ in node i of zone z. Clearly, the variables related to the number of installations in node i of the distribution grid in zone z are upper bounded by the following set of constraints (where $n_i^{\text{CS,max}}$ is the maximum value of allowable CS):

$$\sum_{\psi \in \Psi} n_{\psi,i} \leq n_i^{\text{CS,max}} \quad i \in N_z, z \in \tilde{Z}. \tag{9.2}$$

The second kind of constraint introduces the satisfaction of the power demand in each zone z by the CS. We have used an inequality (instead of equality) constraint to avoid possible infeasibilities. This modeling is also conservative from an engineering point of view. That is:

$$\sum_{i \in N_z} P_{i,m,t}^{\text{CS}} \geq D_{z,m,t}^{\text{EV}} \quad z \in \tilde{Z}, m \in \tilde{M}, t \in \tilde{T} \tag{9.3}$$

where $D_{z,m,t}^{\text{EV}}$ is the overall power demand relevant to EVs in the zone z of month m in any time step t, which is a given parameter for the proposed optimization problem; this data is a priori known and comes from an estimation of the charging demand for

EVs in a given geographic area that includes one or more nodes in the distribution network.

The overall maintenance cost (expressed as a function of the installed active power) of the CS in node i of zone z of month m, is given by the following linear relationship:

$$C^{\mathrm{CS}}_{i,m,t} = \mu^{\mathrm{CS}}_{i,m} P^{\mathrm{CS}}_{i,m,t} \Delta t \quad i \in \tilde{N}, m \in \tilde{M}, t \in \tilde{T} \tag{9.4}$$

where $\mu^{\mathrm{CS}}_{i,m}$ is the unit maintenance cost and Δt is the time step.

9.3.2 Renewable Generators Modeling

In this subsection, the models for each generator to be installed are considered. The solar power plants considered in the model are described by the following equations:

$$P^{\mathrm{PV}}_{i,m,t} = p^{\mathrm{PV}}_{i,m,t} S^{\mathrm{PV}}_{i} \quad i \in N, m \in \tilde{M}, t \in \tilde{T} \tag{9.5}$$

where, $p^{\mathrm{PV}}_{i,m,t}$ is the rated power per square meter to be installed in node i (that is a given data for our problem), and S^{PV}_{i} is a continuous variable that represents the installed surface. For the sake of simplicity, only the surface of the PV plants is a matter of decision at each node of the distribution network. The surface that can be installed is limited for each node (by the parameter $S^{\mathrm{PV,max}}_{i}$) as follows:

$$0 \le S^{\mathrm{PV}}_{i} \le S^{\mathrm{PV,max}}_{i} \quad i \in N. \tag{9.6}$$

Similarly, the power generated by the wind turbines is represented by

$$P^{\mathrm{WT}}_{i,m,t} = n^{\mathrm{WT}}_{i} p^{\mathrm{WT}}_{i,m,t} \quad i \in N, m \in \tilde{M}, t \in \tilde{T} \tag{9.7}$$

where $p^{\mathrm{WT}}_{i,m,t}$ is the rated power of a single wind turbine installed at node i (for our problem it is given parameter) in month m, and n^{WT}_{i} is an integer decision variable that defines the number of turbines installed at node i of the power network. Obviously, the number of WT installations in node i of the power network is upper bounded by the following set of constraints (where $n^{\mathrm{WT,max}}_{i}$ the maximum value of allowable WT):

$$n^{\mathrm{WT}}_{i} \le n^{\mathrm{WT,max}}_{i} \quad i \in N. \tag{9.8}$$

Finally, as regards CHP, the total active power generated in a certain node i can be represented as:

$$P^{\mathrm{CHP}}_{i,m,t} = n^{\mathrm{CHP}}_{i} p^{\mathrm{CHP}}_{i,m,t} \quad i \in N, m \in \tilde{M}, t \in \tilde{T} \tag{9.9}$$

$$n_i^{\mathrm{CHP}} \leq n_i^{\mathrm{CHP,max}} \quad i \in N \tag{9.10}$$

where the variables have the same meaning as the WT case.

In the following, the formulation for the operational costs (i.e., energy costs over the entire optimization period) of each renewable generation plant is given. The costs of photovoltaic plants and wind turbines are related only to the maintenance since the power production is uniquely related to the atmospheric conditions (wind power and solar irradiance):

$$C_{i,m,t}^{\mathrm{PV}} = \mu_{i,m}^{\mathrm{PV}} P_{i,m,t}^{\mathrm{PV}} \Delta t \quad i \in N, m \in \tilde{M}, t \in \tilde{T} \tag{9.11}$$

$$C_{i,m,t}^{\mathrm{WT}} = \mu_{i,m}^{\mathrm{WT}} P_{i,m,t}^{\mathrm{WT}} \Delta t \quad i \in N, m \in \tilde{M}, t \in \tilde{T} \tag{9.12}$$

where $\mu_{i,m}^{\mathrm{PV}}$ and $\mu_{i,m}^{\mathrm{WT}}$ are unit maintenance costs per unit of energy.

The CHP units' costs instead need also to include the amount of fuel needed to feed the entire plant (e.g., natural gas or fossil fuels) and are determined by:

$$C_{i,m,t}^{\mathrm{CHP}} = P_{i,m,t}^{\mathrm{CHP}} \left(\frac{c_{i,m}^{\mathrm{CHP}}}{\eta_{m,t}^{\mathrm{CHP}}} + \mu_{i,m}^{\mathrm{CHP}} \right) \Delta t \quad i \in N, m \in \tilde{M}, t \in \tilde{T} \tag{9.13}$$

where $c_{s,y,m}^{\mathrm{CHP}}$ represents CHPs fuel unit cost, $\eta_{m,t}^{\mathrm{CHP}}$ is the efficiency of the CHP unit, and $\mu_{i,m}^{\mathrm{CHP}}$ is the unit maintenance cost per unit of energy produced by the CHP.

9.3.3 *Power Network Modeling*

The power network modeling is characterized by two sets of equations: the first one is related to the model of the active and reactive nodal power balances (9.14) and (9.15), while the second one models the power flows on each branch of the power network (9.22) and (9.23). The power balance that must be satisfied at each node i is given by:

$$P_{i,m,t}^{\mathrm{PV}} + P_{i,m,t}^{\mathrm{CHP}} + P_{i,m,t}^{\mathrm{WT}} - P_{i,m,t}^{\mathrm{CS}} - P_{i,m,t}^{D} = \sum_{j \in N} P_{i,j,m,t} \quad i \in N, m \in \tilde{M}, t \in \tilde{T} \tag{9.14}$$

$$Q_{i,m,t}^{\mathrm{PV}} + Q_{i,m,t}^{\mathrm{CHP}} + Q_{i,m,t}^{\mathrm{WT}} - Q_{i,m,t}^{\mathrm{CS}} - Q_{i,m,t}^{D} = \sum_{j \in N} Q_{i,j,m,t} \quad i \in N, m \in \tilde{M}, t \in \tilde{T} \tag{9.15}$$

where $P_{i,j,m,t}$ and $Q_{i,j,m,t}$ are the power and the active and reactive power flows between the nodes i and j of the network, and $P_{i,m,t}^{D}$ is the local active power demand in node i (that is a known input data for our problem). Instead $Q_{i,m,t}^{\mathrm{PV}}$, $Q_{i,m,t}^{\mathrm{CHP}}$ $Q_{i,m,t}^{\mathrm{WT}}$ $Q_{i,m,t}^{\mathrm{CS}}$

are the respective reactive power generation for each plant of the node, while $Q^D_{i,m,t}$ is the local reactive power demand (that is a known input data for our problem). The relation that couples active and reactive of each plant is proportional to the power factor $\cos(\varphi)$ as follows:

$$Q^{\mathrm{PV}}_{i,m,t} = P^{\mathrm{PV}}_{i,m,t} \tan(\cos(\varphi^{PV})^{-1}) \quad i \in N, m \in \tilde{M}, t \in \tilde{T} \tag{9.16}$$

$$Q^{\mathrm{CHP}}_{i,m,t} = P^{\mathrm{CHP}}_{i,m,t} \tan(\cos(\varphi^{\mathrm{CHP}})^{-1}) \quad i \in N, m \in \tilde{M}, t \in \tilde{T} \tag{9.17}$$

$$Q^{\mathrm{WT}}_{i,m,t} = P^{\mathrm{WT}}_{i,m,t} \tan(\cos(\varphi^{\mathrm{WT}})^{-1}) \quad i \in N, m \in \tilde{M}, t \in \tilde{T} \tag{9.18}$$

$$Q^{\mathrm{CS}}_{i,m,t} = P^{\mathrm{CS}}_{i,m,t} \tan(\cos(\varphi^{\mathrm{CS}})^{-1}) \quad i \in N, m \in \tilde{M}, t \in \tilde{T}. \tag{9.19}$$

To model the capacity of every secondary substation we introduce the following inequality constraints to limit the nodal injection of active and reactive power:

$$P^{\mathrm{min}}_i \le \sum_{j \in N} P_{i,j,m,t} \le P^{\mathrm{max}}_i \quad i \in N, m \in \tilde{M}, t \in \tilde{T} \tag{9.20}$$

$$Q^{\mathrm{min}}_i \le \sum_{j \in N} Q_{i,j,m,t} \le Q^{\mathrm{max}}_i \quad i \in N, m \in \tilde{M}, t \in \tilde{T}. \tag{9.21}$$

The electrical network is represented by means of the power flow Eqs. (9.22) and (9.23) which describe the sinusoidal steady-state equilibrium of a power network, formulated regarding Cartesian coordinates representation of circuit quantities. That is:

$$\begin{aligned} p_{i,j,m,t} = {} & G_{i,j}(v_{i,m,t})^2 - v_{i,m,t}v_{j,m,t}G_{i,j}\cos(\delta_{i,m,t} - \delta_{j,m,t}) \\ & + v_{i,m,t}v_{j,m,t}B_{i,j}\sin(\delta_{i,m,t} - \delta_{j,m,t}) \quad i, j \in N, m \in \tilde{M}, t \in \tilde{T} \end{aligned} \tag{9.22}$$

$$\begin{aligned} q_{i,j,m,t} = {} & -B_{i,j}(v_{i,m,t})^2 + v_{i,m,t}v_{j,m,t}B_{i,j}\cos(\delta_{i,m,t} - \delta_{j,m,t}) \\ & - v_{i,m,t}v_{j,m,t}G_{i,j}\sin(\delta_{i,m,t} - \delta_{j,m,t}) \quad i, j \in N, m \in \tilde{M}, t \in \tilde{T} \end{aligned} \tag{9.23}$$

where all the values are in p.u. and, in particular, $p_{i,j,m,t}$ and $q_{i,j,m,t}$ are the active and reactive power flows between the nodes i and j, $v_{i,m,t}$ and $\delta_{i,m,t}$ are the voltage and the phase of node i, $G_{i,j}$, and $B_{i,j}$ are the conductance and susceptance matrices.

Finally, power quality requirements from standards necessitate the introduction of some upper and lower bounds on the voltage magnitude, phase, and power flow on each branch of the power distribution network:

$$\delta^{\mathrm{min}} \le \delta_{i,m,t} \le \delta^{\mathrm{max}} \quad i \in N, m \in \tilde{M}, t \in \tilde{T} \tag{9.24}$$

$$v^{\min} \leq v_{i,m,t} \leq v^{\max} \quad i \in N, m \in \tilde{M}, t \in \tilde{T} \tag{9.25}$$

$$p_{i,j}^{\min} \leq p_{i,j,m,t} \leq p_{i,j}^{\max} \quad i,j \in N, m \in \tilde{M}, t \in \tilde{T}. \tag{9.26}$$

9.3.4 Objective Function

The objective function to be minimized is related to net present value of the annual costs plus the installation cost of each plant, given by:

$$\min\left\{K + \sum_{y=1}^{Y} \frac{C_y}{(1+\chi)^y}\right\} \tag{9.27}$$

where K is the cost for the installation of the power plants, C_y is their annual cost, and χ the discount rate. In particular, K is defined as a linear combination of the installation variables of the plants as follows:

$$K = \sum_{i \in \tilde{N}} \left(k^{\mathrm{PV}} S_i^{\mathrm{PV}} + k^{\mathrm{CHP}} n_i^{\mathrm{CHP}} + k^{\mathrm{WT}} n_i^{\mathrm{WT}} + k^{\mathrm{CS}} n_i^{\mathrm{CS}}\right) \tag{9.28}$$

where k^{PV}, k^{CHP}, k^{WT} and k^{CS} are the installation costs for every installed unit. Note that if no plant is installed, K is equal to 0.

The annual costs C_y are determined as direct sum of the plant cost and the energy purchased/bought from/to the external power grid, that is determined by:

$$C_y = \left[\sum_{i \in N} \sum_{m \in \tilde{M}} \sum_{t \in \tilde{T}} n_m^{\mathrm{days}} \left(C_{i,m,t}^{\mathrm{GRID}} + C_{i,m,t}^{\mathrm{CHP}} + C_{i,m,t}^{\mathrm{PV}} + C_{i,m,t}^{\mathrm{WT}} + C_{i,m,t}^{\mathrm{CS}}\right)\right] \quad y = 1, \ldots, Y \tag{9.29}$$

with

$$C_{i,m,t}^{\mathrm{GRID}} = c_{i,m,t}^{\mathrm{el}} E_{i,m,t}^{\mathrm{IN}} - b_{i,m,t}^{\mathrm{el}} E_{i,m,t}^{\mathrm{OUT}} \quad i \in N, m \in \tilde{M}, t \in \tilde{T} \tag{9.30}$$

$$E_{i,m,t}^{\mathrm{IN}} = \max\left(0, \sum_{j \in N} P_{i,j,m,t} \Delta t\right) \quad i \in N, m \in \tilde{M}, t \in \tilde{T} \tag{9.31}$$

$$E_{i,m,t}^{\mathrm{OUT}} = \min\left(0, \sum_{j \in N} P_{i,j,m,t} \Delta t\right) \quad i \in N, m \in \tilde{M}, t \in \tilde{T} \tag{9.32}$$

where $c^{el}_{i,m,t}$ and $b^{el}_{i,m,t}$, are the price for selling and purchasing electricity from the grid, respectively.

The overall optimization problem is given by the minimization in Eq. (9.27) subject to constraints (9.1)–(9.26) and (9.28)–(9.32).

9.4 Case Study Application

The model described in the previous section gives rise to a Mixed-Integer Nonlinear optimization Problem (MINLP) that has been solved through the LINGO [13] optimization tool. The overall optimization has been carried out over a prediction horizon of one year.

The proposed planning model has been tested on a modified IEEE 13 bus network (reported in Fig. 9.2), representing a district with four zones in which CSs can be installed. For brevity, we consider a single-phase version of the network, even though an extension to a three-phase network can be done trivially.

The optimization problem (9.1)–(9.32) requires an appropriate amount of data for each node and zone, which can be withdrawn from historical data or available databases and models. Specific measures of renewable sources may commonly be

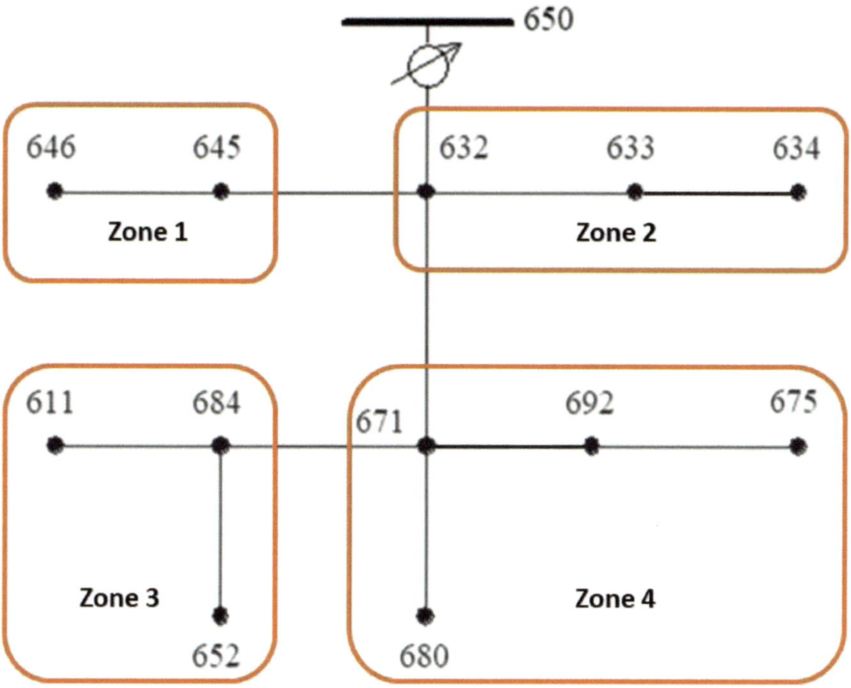

Fig. 9.2 Modified IEEE 13 bus network

unavailable, and the necessary choice of the measures to be considered leads to uncertainties in the results. As far as the technologies are concerned, we consider polycrystalline PV fields, WT with 15 kW of rated power, and CHP units with a rated electrical power equal to 35 kW with full load efficiency of 26%. Finally, two types of CSs can be chosen, i.e., 22 and 50 kW, the most common types of commercial technologies. The buying price is 0.15 [€/kWh], and the selling price is 0.08 [€/kWh].

In Figs. 9.3, 9.4 and 9.5, the wind power availability, the PV power availability, and the total energy demand profiles for each zone of the case study are reported. Four representative nodes (one for each zone) are shown as regards wind and PV.

Table 9.1 shows the optimal results for the considered case study, reporting for each node of the distribution grid (except for node 650 that is the primary substation), the number of WTs, CHPs, the size of the PV, and the placement of CSs. The value of the objective function is 325.347 k€ with a run time of about 90 s.

According to the results, distributed generation is convenient when the produced electricity is self-consumed because the electricity buying price, even in off-peak hours, is always considerably higher than the selling price. Regarding the CSs' location, since the voltage magnitude limits are considered constraints of the optimization problem, the charging stations are placed into nodes where the voltage drop is less affected, i.e., not at the end of the lines. For example, this is evident in Zone 4 at node 675, where no CS is installed. Otherwise, a large amount of PV is placed. This compensates for the voltage drop caused by the installation of CS in the same zone. Similar considerations can be made for the other nodes of the distribution network.

Fig. 9.3 Wind power on the different network zones

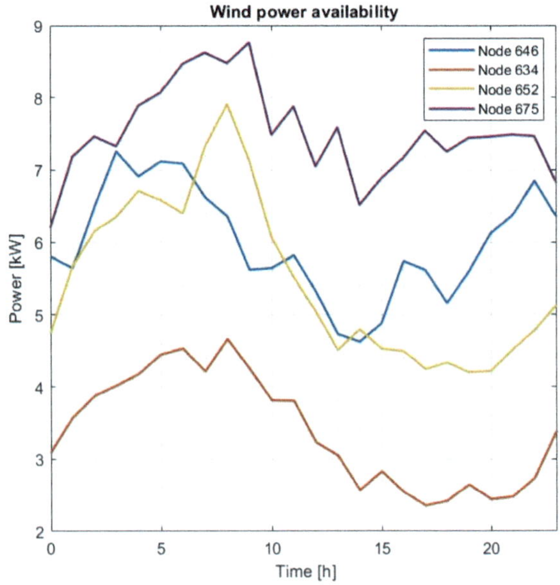

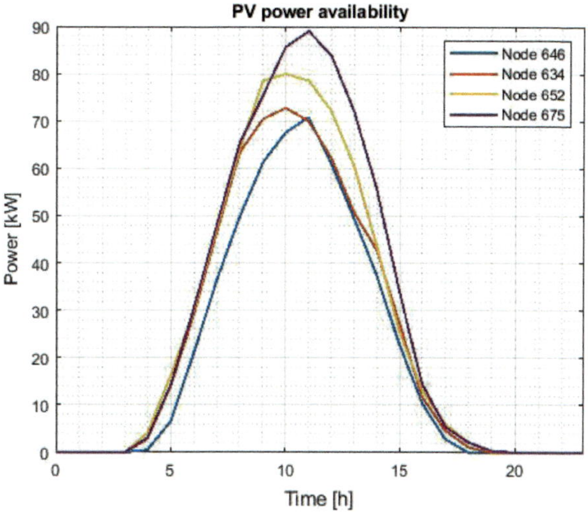

Fig. 9.4 PV power on the different network zones

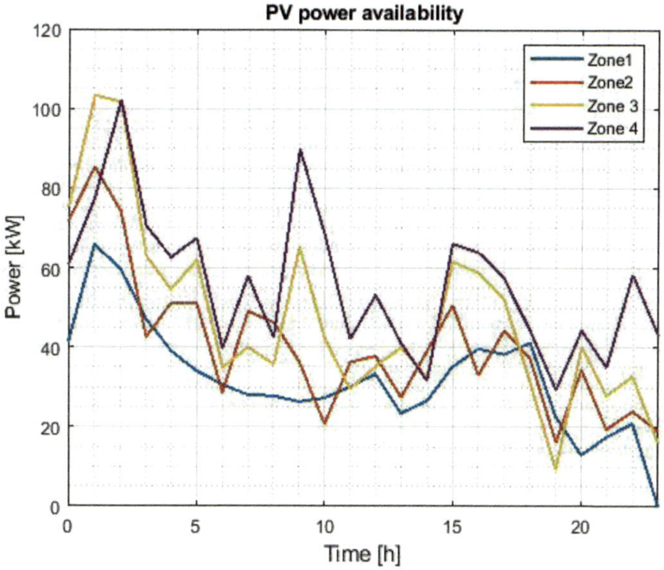

Fig. 9.5 EVs demand in the different network zones

Table 9.1 Optimal placement results

Node number	#WT	#CHP	#CS 22 kW	#CS 50 kW	PV [m^2]
645	2	1	2	0	120
646	0	3	1	0	135
632	0	1	0	1	150
633	1	1	2	0	129
634	0	1	0	0	86
611	1	1	0	0	149
684	0	2	0	2	137
652	0	1	1	0	148
671	2	1	1	0	168
680	0	1	1	0	25
692	2	1	0	1	195
675	0	1	0	0	203

References

1. Bracco S, Delfino F, Rossi M, Robba M, Pagnini L (2016) Optimal planning of the energy production mix in smart districts including renewable and cogeneration power plants. In: 2016 IEEE international smart cities conference (ISC2), pp 1–7
2. Clercq SD, Zwaenepoel B, Vandevelde L (2018) Optimal sizing of an industrial microgrid considering socio-organizational aspects. IET Gener Transm Distrib 12(14):3442–3451
3. Haghi HV, Lotfifard S et al (2015) Spatiotemporal modeling of wind generation for optimal energy storage sizing. IEEE Trans Sustain Energy 6(1):113–121
4. Xing H, Fan H, Sun X et al (2018) Optimal siting and sizing of distributed renewable energy in an active distribution network. CSEE J Power Energy Syst 4(3):380–387
5. Saha S, Mukherjee V (2016) Optimal placement and sizing of DGs in RDS using chaos embedded SOS algorithm. IET Gener Transm Distrib 10(14):3671–3680
6. Ferrari L, Bianchini A, Galli G (2018) Influence of actual component characteristics on the optimal energy mix of a photovoltaic-wind-diesel hybrid system for a remote off-grid application. J Clean Prod 178:206–219
7. Singh S, Fernandez E (2018) Modeling size optimization and sensitivity analysis of a remote hybrid renewable energy system. Energy 143:719–731
8. Shi Z, Wang R, Zhang T (2015) Multi-objective optimal design of hybrid renewable energy systems using preference-inspired coevolutionary approach. Sol Energy 118:96–106
9. Erdinc O, Tascikaraoglu A, Paterakis NG et al (2017) Comprehensive optimization model for sizing and siting of DG units, EV charging stations, and energy storage systems. IEEE Trans Smart Grid 9(4):3871–3882
10. Masaud TM, Oyebanjo O, Sen P (2017) Sizing of large-scale battery storage for off-grid wind power plant considering a flexible wind supply-demand balance. IET Renew Power Gener 11(13):1625–1632
11. Gao X et al (2020) Optimization of EV-fast charging station placement for grid support. IFAC-PapersOnLine 53(2):13769–13774

12. Phonrattanasak P, Leeprechanon N (2014) Optimal placement of EV fast charging stations considering the impact on electrical distribution and traffic condition. In: 2014 international conference and utility exhibition on green energy for sustainable development (ICUE). IEEE, pp 1–6
13. Lingo optimization suite. https://www.lindo.com/index.php/products/lingo-and-optimization-modeling

Chapter 10
Conclusions

The progressive shift from traditional vehicles to EVs (Electric Vehicles) is considered as one of the key measures to achieve the objective of a significant reduction in the emission of pollutants, especially in urban areas. In this book, attention has been focused on applying control and optimization methods and approaches to energy systems in which EVs (Electric Vehicles) are present, with specific reference to planning and scheduling decision problems. An interdisciplinary approach is considered because the optimal management of EVs and charging stations (CSs) is a decision problem across different disciplines: transportation and logistics, scheduling, and smart grids. In smart grids, energy production and storage systems are usually scheduled by an Energy Management System (EMS) to minimize costs, power losses, and CO_2 emissions while satisfying energy demands. When CSs are connected to a smart grid, EVs served by CSs represent an additional load to the power system to be satisfied and an additional storage system in the case vehicle-to-grid technology is enabled. However, the load generated by EVs is deferrable. In this case, it can be thought of as a process in which there are machines (CSs) that serve customers/products (EVs) based on release time, due date, deadline, and energy request, like it happens in manufacturing systems. In this book, two kinds of approaches have been used to deal with the optimal scheduling of EVs. Firstly, attention is focused on the formalization of a discrete-time optimization problem in which fossil fuel production plants, storage systems, and renewables are considered to satisfy the grid's electrical load. The discrete-time formalization can use forecasting for renewables and loads without data elaboration. On the other side, a huge number of decision variables are present, making the optimization problem hard to solve through commercial optimization tools. Secondly, the same decision problems have been faced through a discrete-event formalization. Both public and private EVs have been considered.

G. Ferro et al., *Optimization of Electric-Vehicle Charging*, Advances in Industrial Control, https://doi.org/10.1007/978-3-031-61917-5_10

However, when dealing with EVs, it is necessary to consider other issues that influence the scheduling, i.e., the transportation demand assessment, the optimal location of charging stations, and the assignment of users to CSs. For this reason, we have dedicated different chapters to planning problems and charging station placement, taking into account the traffic network and User Equilibrium conditions both in the deterministic and stochastic cases.

In particular, the book has been structured in the following way. Chapter 1 is an introduction to smart grids and the characteristics of EVs and CSs. The role of modeling, simulation, and optimization is then highlighted in Chap. 2, in which the state of the art and the current challenges will be described. From Chaps. 3 to 6, the book is focused on the optimal scheduling of electric vehicles with different possible options in the problem formalization: discrete-time and discrete-event optimization, a-periodic and periodic decision models. The reason lies in the fact that the different approaches want to be described in detail in order to give the basis for the development of models for the scheduling of EVs. In particular, Chap. 3 focuses attention on the scheduling of EVs through a discrete-time formalization. The proposed model is deterministic, nonlinear, and has continuous and binary variables. It is based on forecasting for demands and renewable power availability. Chapter 4 presents a novel discrete-event optimization model; this approach has the characteristic of reducing the number of decision variables and decreasing the run time. In particular, attention is paid to the nonlinear behavior of the battery.

Even Chaps. 5 and 6 are based on a discrete-event formalization. Chapter 5 presents an extension of the model presented in Chap. 4. Here the charging station is equipped with more sockets, allowing the simultaneous charging of a chosen number of electric vehicles (EVs). The model is tested on a case study consisting of different scenarios with variable energy demand. Chapter 6 focuses on electric buses (EBs). Even if this kind of vehicle can be considered analogous to private electric vehicles, some differences deserve to be pointed out. In particular, EBs are usually characterized by well-defined time patterns, leading to particular hypotheses on their scheduling. For this reason, a discrete-event optimization model for periodic scheduling is proposed. The presented case study is defined starting from some real data provided by AMT, the bus company operating in Genoa municipality.

The remaining chapters are dedicated to planning problems. In particular, Chap. 7 deals with user equilibrium models for energy demand assessment and traffic assignment. Specifically, the so-called stochastic user equilibrium (SUE) conditions are used to define the steady-state behavior of a generic traffic network where some arcs are equipped with EV charging stations. SUE conditions under different user's path choice functions have been described, namely the multinomial logit and the multinomial C-logit. Chapter 8 is dedicated to a bi-level approach that at the lower level is characterized by user equilibrium conditions and at the higher level the sizing and location of CSs are formalized. In particular, a bi-level optimization problem uses the traffic model to derive the SUE, based on a stochastic choice model derived from the Logit model. The possible paths are determined on the basis of present charging stations and new potential charging facilities. Then, in Chap. 9, the detailed placement on a distribution grid of CSs is discussed, which has, as an input, the results of the analysis reported in Chaps. 7 and 8, and provides the optimal results for location and sizing of CSs, taking into account the electrical constraints of a smart grid and its EMS. The considered technologies are small-size WTs, PV, and cogeneration micro-turbines connected to the same grid, and the overall net present value is minimized. The optimal results are obtained over a modified IEEE 13 bus network.

Finally, an Appendix is introduced to remind concepts of optimization and modeling.

Of course, other topics should be mentioned that are worthy of investigation in the planning and management of EVs; among them:

- *Decision problems related to car, scooter, and bike-sharing.* EVs are prime candidates for use in urban car-sharing systems. However, several decision problems arise that can be resolved using optimization models. These include determining the best locations and sizes for charging stations, establishing the maximum number of regular trips, repositioning vehicles in free-floating electric car-sharing systems, swapping batteries, managing stations, picking up and delivering vehicles, and scheduling optimal charging and swapping operations.
- *Routing and Charging of electric vehicles.* Electric vehicles are used to transport goods (freight transportation services). People have to define the optimal route to a destination and where and how much to charge, taking into account the characteristics of the available infrastructure of charging stations (i.e., location, type, and electrical constraints), charging times, energy consumption, and cost. This problem extends the well-known vehicle routing problem with time windows (VRPTW), and it belongs to the class of Green VRP (GVRP). From the computational standpoint, Electric VRP (EVRP) is NP-hard since it generalizes the classic VRP, which has been demonstrated to be NP-hard. Only a few works in the very recent literature refer to the routing of EVs, including time-varying energy prices, battery consumption, type and behavior of charging stations, and maximum available power.
- *Power management in charging stations and smart parkings.* Power management in charging stations involves utilizing control algorithms and rules for enabling power modulation among electric vehicles connected to multiple sockets of a charging station or different charging stations. Typically, these control algorithms and rules are integrated within the charging station and operate in real-time upon the presence of a vehicle starting to charge, thereby guaranteeing compliance with the maximum power that can be sourced from the external grid. This is a common scenario at a charging park where information about privately owned electric vehicles that arrive at the charging station is lacking. The main problems are related to the state-of-charge and state-of-health identification, control algorithms embedded in CSs, management, and control of EVs in smart parking.

It should be noted that the regulation related to energy systems and markets is evolving, and with it, the necessary technologies and tools. Some specific examples are the market aggregator and the energy communities (ECs).

A market aggregator is an economic entity that operates in the balancing market and that has the responsibility of coordinating different customers in order to reduce the electrical load of a certain area, as a consequence of the requests of the Transmission System Operator. An aggregator, in order to receive the remuneration for load reduction has to define policies and incentives that take into account the flexibility capabilities of local prosumers and their costs. In the recent literature, aggregators of EVs are emerging. In fact, there are some kinds of EVs (i.e., electrical buses) that can provide a huge reserve of flexibility (since they act as storage systems). Moreover, charging parks of many EVs can act as prosumers in the energy market adopting smart charging and vehicle-to-grid policies. It is important to know that many challenges have to be addressed in this framework. First of all, it is necessary to take into account standards for parameters communication and constraints coming from manufacturers. Then, from a modeling point of view, in order to include EVs' charging parks in balancing market optimization problems, it is necessary to consider the dynamic behavior, thus increasing the complexity and the runtime in the solution of the decision models.

Significant efforts have recently been made worldwide to decrease greenhouse gas emissions and achieve sustainability, resulting in a profound shift in the management of resources and cities. At the European level, the European Green Deal specifically addresses climate change and environmental degradation by decreasing greenhouse gas emissions, promoting sustainable economic growth, and focusing on citizens. ECs present a substantial opportunity to utilize sustainable energy sources, enhance energy efficiency in edifices, transit, and manufacturing procedures, and embolden the role of citizens in the energy marketplace through active participation made plausible by fresh regulations and emerging players. In particular, adopting ECs can contribute to several benefits of the European Green Deal. These include energy-efficient buildings, cleaner energy, cutting-edge clean technological innovations, future-proof jobs, skills training for the transition, and globally competitive and resilient industries. These are included in the following main actions: climate, energy, research, and innovation.

An EC refers to a group of residential or small commercial entities, each functioning as a prosumer and typically consisting of electric and thermal generation, battery storage units, and flexible loads. The European Union's Clean Energy Package identifies specific types of EC initiatives. 'Citizen energy communities' and 'renewable energy communities' are both included in the revised Internal Electricity Market Directive 2019/944 (European Parliament & Council of the European Union, 2019) and the revised Renewable Energy Directive 2018/2001 (European Parliament & Council of the European Union, 2018). A notable characteristic of ECs involves the collaborative utilization of individual resources, like PV panels, batteries, and CHP (Combined Heat and Power) units, among others, for the community's mutual benefit. The cooperative use of resources in the community reflects a vital benefit of ECs. ECs can also share "common" resources, such as renewable energy plants, batteries, hydrogen production, storage and re-electrification facilities, EVs' charging areas, and thermal units. Moreover, ECs can also be seen as local prosumers that, through the figure of an aggregator, can help in demand response programs (i.e., shifting and reducing load in a certain area and time period). In order to carry out optimal management of these distributed resources, new solutions consisting of automation, digitalization, ICT platforms, and IoT (Internet of Things) should be developed.

From a modeling and optimization point of view, it is necessary to develop:

- optimization models for operational management that consider shared resources;
- market architectures within a local area with different prosumers and coupled networks;
- tools for the optimal design of ECs and multiple ECs over a territory;
- dynamic optimization problems for the inclusion of ECs in the energy market.

Another key priority for EU is digitalization, and many programs are devoted to digital transformation of the energy system (https://digital-strategy.ec.europa.eu/en/policies/eu-programmes-digitalisation-energy). In this framework, the role of ICT platforms, automation, machine learning, and cyber security assume a central role. From a modeling and optimization point of view, this means developing energy management and building automation systems able to operate in real-time, considering stochastic processes, including resiliency to cyber-attacks and natural risks, and using distributed optimization.

Finally, it should be noted that EVs not only influence the management of power and energy systems but also the one of sustainable transportation and mobility systems. This is the reason why it is to couple different networks and integrate different disciplines such as traffic systems, autonomous vehicles, electrical systems, transportation networks, and logistics.

Appendix A[1]

A.1 Optimization Problems

An optimization problem is characterized by three main components: decision variables, objective function, and constraints. Decision variables are those entities that represent the decisions of a specific decision problem (for example power produced by some production plant, the number and the size of charging stations in a territory, the path of a vehicle, etc.). The objective function represents the target (or the key performance indicators) of the decision problem to be minimized or maximized (like for example costs, revenues, emissions, temperature, and state-of-charge levels); it must be a function of the decision variables and it is generally given by an analytic representation. Constraints represent restrictions affecting the choice of the decision variables, for instance as regards the size of production plants, the emission levels (on the basis of specific regulations), etc. An optimization problem is generally represented as follows:

$$\min_{x} f(x) \tag{A.1}$$

s.t.

$$h(x) = 0 \tag{A.2}$$

$$g(x) \le 0 \tag{A.3}$$

[1] The work in this Section is © 2009 Springer-Verlag US, reproduced from [1], https://doi.org/10.1007/978-0-387-74759-0; with permission.

G. Ferro et al., *Optimization of Electric-Vehicle Charging*, Advances in Industrial Control, https://doi.org/10.1007/978-3-031-61917-5

where $x \in \mathbb{R}^n$ is the decision variable vector, $f(x): \mathbb{R}^n \to \mathbb{R}$ is the objective function to be minimized, $h(x): \mathbb{R}^n \to \mathbb{R}^{n_E}$ are the functions for the equality constraints and $g(x): \mathbb{R}^n \to \mathbb{R}^{n_I}$ for the inequality constraints.

It is important to note that:

- A feasible solution is any $x \in \mathbb{R}^n$ that respects constraints (A.2) and (A.3).
- If a feasible solution minimizes (A.1), it is an optimal solution, which then provides the objective function optimal value.
- Minimizing $f(x)$ is equivalent to maximizing $-f(x)$.

Example: Linear Optimization Problem A linear programming (LP) problem is a mathematical programming problem in which functions f, $\underline{g}$, and $\underline{h}$ are all linear functions (or affine functions) of the vector $\underline{x}$ of the *decision variables*. An affine function is a linear function plus a constant.

Any LP problem can be put in the *standard form:*

$$\max \quad x_0 = \underline{c}^T \underline{x} \tag{A.4}$$

s.t.

$$A\underline{x} = \underline{b} \tag{A.5}$$

$$\underline{x} \geq 0 \tag{A.6}$$

where

$$\begin{aligned} &\underline{x} \in R^n \\ &\underline{c}^T \in R^n \\ &\underline{b} \in R^m \\ &\dim[A] = m \cdot n \end{aligned}$$

besides, it is supposed that $m < n$ and that $rank[A] = m$.

Note that $\underline{c}^T$, A, $\underline{b}$ are the known parameters of the optimization problem.

Remark A.1 The justification for the above claim of the generality of the standard form is relatively easy. Any linear inequality constraint may be put into an equality form by introducing additional non-negative auxiliary variables. For example:

$$\begin{aligned} 3x_1 + 2x_2 \leq 5 \quad &\leftrightarrow \quad \begin{cases} 3x_1 + 2x_2 + x_3 = 5 \\ x_3 \geq 0 \end{cases} \\ 3x_1 + 5x_2 \geq 7 \quad &\leftrightarrow \quad \begin{cases} 3x_1 + 5x_2 - x_3 = 7 \\ x_3 \geq 0 \end{cases}. \end{aligned} \tag{A.7}$$

Besides, if a variable is unrestricted in sign, it can be represented via two variables restricted in sign:

$$x_1 \leq\geq 0 \quad \leftrightarrow \quad \begin{cases} x_1 = x_1^+ - x_1^- \\ x_1^+ \geq 0 \\ x_1^- \geq 0 \end{cases} . \tag{A.8}$$

Finally, it should be noted that any minimization problem can be put in the form of a maximization problem and vice-versa:

$$\min\ (3x_1 - 5x_2) \quad \leftrightarrow \quad \max\ (-3x_1 + 5x_2). \tag{A.9}$$

Regarding the standard form of an LP problem,

(a) A vector $\underline{x}$ satisfying the matrix equality $A\underline{x} = \underline{b}$ is said to be a solution;
(b) The solution $\underline{x}$ is said *feasible* if it also satisfies the non-negativity constraint $\underline{x} \geq 0$;
(c) A feasible solution $\underline{x}^*$ is *optimal* if no other feasible solution $\underline{x}'$ exists such that $\underline{c}^T\underline{x}' > \underline{c}^T\underline{x}^*$ (that is if no better feasible solution exists).

A.2 Graph Representation

A simple graph $G(V, E)$ is characterized by vertices (or nodes) V and pairs of distinct elements of V called edges (E). Graphs can be classified into two main categories:

- Directed graph: edges have a direction (Fig. A.1).
- Undirected graph: all the edges are bidirectional (Fig. A.2).

A graph is characterized by the adjacency matrix, which is defined as follows:

$$\begin{cases} A_{ij} = 1, \text{ if } ij \in E \\ A_{ij} = 0, \text{ otherwise} \end{cases}$$

The adjacency matrices of the graphs depicted in Figs. A.1 and A.2 are given by:

$$A_{ud} = \begin{pmatrix} 0\,1\,1\,1\,1 \\ 1\,0\,1\,0\,0 \\ 1\,1\,0\,1\,1 \\ 1\,0\,1\,0\,0 \\ 1\,0\,1\,0\,0 \end{pmatrix} \tag{A.10}$$

Fig. A.1 A directed graph

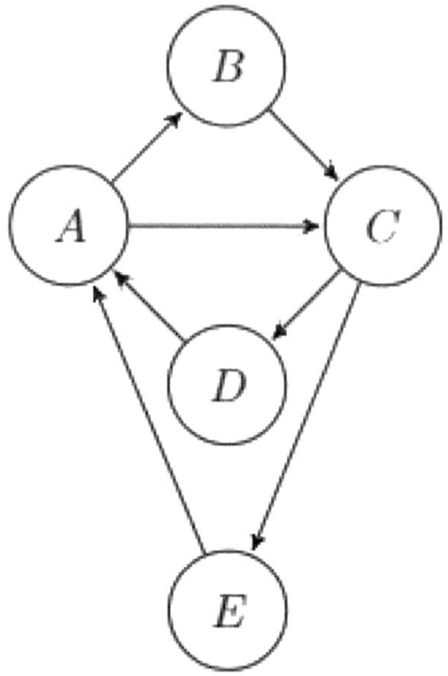

Fig. A.2 An undirected graph

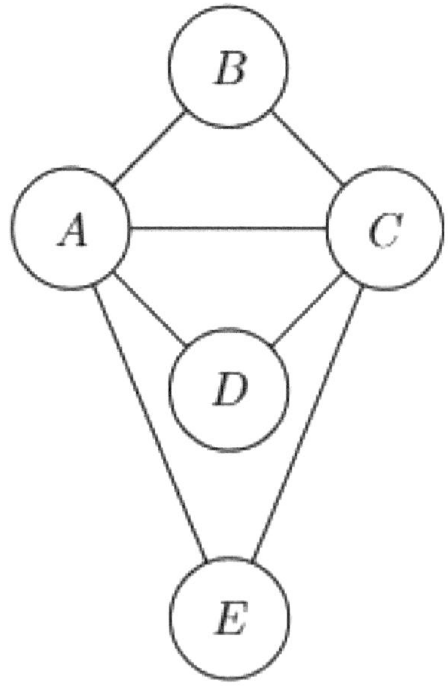

$$A_d = \begin{pmatrix} 0 & 1 & 1 & 0 & 0 \\ 0 & 0 & 1 & 0 & 0 \\ 0 & 0 & 0 & 1 & 1 \\ 1 & 0 & 0 & 0 & 0 \\ 1 & 0 & 0 & 0 & 0 \end{pmatrix}. \tag{A.11}$$

A.3 Modeling a Dynamic System[2]

A dynamic system is characterized by input ($\mathbf{u}(t)$), output ($\mathbf{y}(t)$), and state ($\mathbf{x}(t)$) vectors within an initial time t_0 and a final time t_f [2].

The state of a system at time t_0 is required as initial condition. The components $x_1(t), \ldots, x_n(t)$ are called *state variables*. The set of equations required to specify the state $\boldsymbol{x}(t)$ for all $t \geq t_0$ given $x(t_0)$ and the function $\mathbf{u}(t)$, $t \geq t_0$, are called state equations. The following representation based on differential equations for the system state is used:

$$\dot{\mathbf{x}}(t) = \mathbf{f}(\mathbf{x}(t), \mathbf{u}(t), t). \tag{A.12}$$

By including also the algebraic relation of the output a dynamical system can be represented in state space form as follows:

$$\begin{aligned} \dot{\mathbf{x}}(t) &= \mathbf{f}(\mathbf{x}(t), \mathbf{u}(t), t) \quad x(t_0) = x_0 \\ \mathbf{y}(t) &= \mathbf{g}(\mathbf{x}(t), \mathbf{u}(t), t) \end{aligned}. \tag{A.13}$$

A system is linear when it takes the following form:

$$\begin{aligned} \dot{x}(t) &= \mathbf{A}(t)x(t) + \mathbf{B}(t)u(t) \\ y(t) &= \mathbf{C}(t)x(t) + \mathbf{D}(t)u(t) \end{aligned} \tag{A.14}$$

where **A**, **B**, **C**, and **D** are given matrices.

The system can be also expressed by a discrete-time formalization. That is:

$$\begin{aligned} x(k+1) &= \mathbf{A}(k)x(k) + \mathbf{B}(k)u(k) \\ y(k) &= \mathbf{C}(k)x(k) + \mathbf{D}(k)u(k) \end{aligned} \tag{A.15}$$

where k represents the time counter.

It is important to note that in a discrete-time formalization, time intervals are equal in length and that control variables are assumed to be constant over the length of time. The state equations can be inserted as constraints in a general optimization problem in which the objective function depends on the state and control variables.

Finally, a system can also be represented by a discrete-event approach. In this case, the state and control variables change when an event occurs. Figure A.3 reports this particular case that should be modeled appropriately; this book details the application to EVs' scheduling.

[2] The work in this Section is © 2021 Springer Nature Switzerland AG, reproduced from [2], https://doi.org/10.1007/978-3-030-72274-6; with permission.

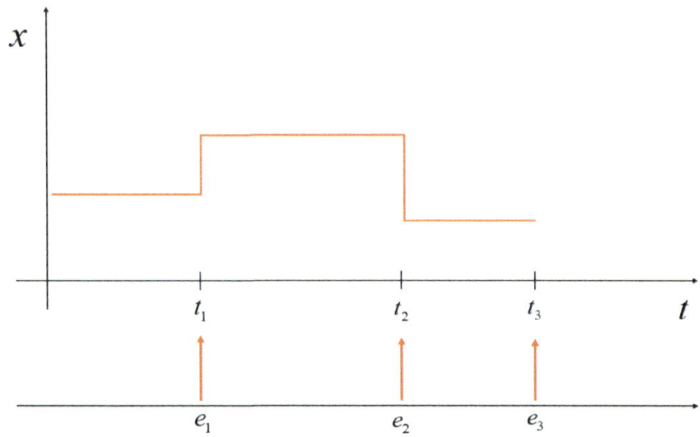

Fig. A.3 Discrete-event case

References

1. Floudas C, Pardalos P (ed) (2008) Encyclopedia of optimization. Springer Science & Business Media
2. Cassandras CG, Lafortune S (2021) Introduction to discrete event systems. Springer, p 822

Index

G. Ferro et al., *Optimization of Electric-Vehicle Charging*, Advances in Industrial Control, https://doi.org/10.1007/978-3-031-61917-5

GPSR Compliance

The European Union's (EU) General Product Safety Regulation (GPSR) is a set of rules that requires consumer products to be safe and our obligations to ensure this.

If you have any concerns about our products, you can contact us on ProductSafety@springernature.com

In case Publisher is established outside the EU, the EU authorized representative is:

Springer Nature Customer Service Center GmbH
Europaplatz 3
69115 Heidelberg, Germany

Batch number: 10732090

Printed by Printforce, the Netherlands